RISK MODELING FOR DETERMINING VALUE AND DECISION MAKING

GLENN KOLLER

CRC Press
Taylor & Francis Group
Boca Raton London New York

CRC Press is an imprint of the
Taylor & Francis Group, an **informa** business

A CHAPMAN & HALL BOOK

CRC Press
Taylor & Francis Group
6000 Broken Sound Parkway NW, Suite 300
Boca Raton, FL 33487-2742

First issued in paperback 2019

ISBN-13: 978-1-58488-167-4 (hbk)
ISBN-13: 978-0-367-39864-4 (pbk)
Library of Congress Card Number 00-022387

Library of Congress Cataloging-in-Publication Data

Koller, Glenn R. (Glenn Robert), 1951-
    Risk modeling for determining value and decision making/ by Glenn R. Koller.
      p. cm.
    Includes index.
  ISBN 1-58488-167-4 (alk. paper)
    1. Risk assessment—Statistical methods. 2. Decision making—Statistical methods.
    3. Value—Statistical methods. I. Title.
HD61 .K633 2000
658.15′5—dc21
                                                      00-022387

Visit the Taylor & Francis Web site at
http://www.taylorandfrancis.com

and the CRC Press Web site at
http://www.crcpress.com

About the Author

Glenn R. Koller has responsibility for aspects of risk and uncertainty analysis, management, and training in BP Amoco (BPA) Corporation. He received his Ph.D. in geochemistry/geophysics from Syracuse University. Responsibilities of his current position include implementation of risk-assessment and risk-management technologies and processes in BPA; development of statistical routines that comprise BPA's risk systems; keeping abreast of risk technology developed by vendors, other companies, and the National Laboratories; marketing risk technology; and performing technical and consulting services. Areas of responsibility include business and product development, environmental concerns, ranking and prioritization of projects and products, and other aspects of BPA's diversified business. Dr. Koller currently resides in Tulsa, Oklahoma. His email address is grk1000@aol.com.

Acknowledgments

I am grateful beyond measure to Mike A. Long and Phil Hammond for their continued friendship, support, and dedication to our risk/uncertainty enterprise. Their teaching, programming, and customer-relations skills are key elements of our successful effort.

I acknowledge and thank Chris Mottershead, BP Amoco's Chief of Staff–Technology, for his review of this volume and permission to publish. His understanding, advice, and guidance are truly appreciated.

I thank Roger Rowe for his encouragement and friendship. I look forward to working with Roger on future projects.

I express my deepest appreciation to my wife Karen, my daughters Stephanie and Kathy, and my son Matthew. Writing a book is an exceedingly time-consuming undertaking. Every hour spent on the book-writing effort is an hour not spent with the family. I truly appreciate their love, patience, and understanding.

To Karen, Stephanie, Kathy, and Matthew —
my reasons for living.
Thank you for your love, patience, and understanding.

Contents

FUNDAMENTALS OF RISK ASSESSMENT

Introduction

CONTENTS

SCOPE

This book is a follow-on to an initial volume entitled *Risk Assessment and Decision Making in Business and Industry: A Practical Guide* published by CRC Press (ISBN 0-8493-0268-4) which henceforth will be referred to as "the first book." In the first book were delineated in narrative style and in plain English the fundamental processes and technologies that compose the art and science of risk assessment and decision making and a few examples thereof. It is assumed that the reader of this volume is familiar with those processes and principles, and much of the material presented in the first book will not be repeated or reviewed in this treatise. Topics covered in the first book include

- Risk assessment process
- Organizational/cultural issues
- Risk communication
- Education
- Risk vision
- Building a consensus model
- Consistency
- Building a contributing-factor diagram
- Double dipping
- Bayesian analysis
- Decision trees
- Factor analysis
- Neural nets
- Monte Carlo analysis
- Distributions
- Decisions
- Chance of failure (abject and financial)
- Time-series analysis
- Dependence
- Risk-weighted values

- Sensitivity analysis
- A host of risk assessment examples

To gain a more complete and comprehensive understanding of the many processes and techniques alluded to in this book (and much more), it is recommended that the reader review the information in the first book. Chapters 5, 6, and 9 through 13 of the first book are, with some modification, repeated in this volume as Chapters 12 through 18 in the Fundamentals of Risk Assessment section. The publisher felt that these chapters were essential to the understanding of the precepts and technologies utilized in the examples of risk and uncertainty assessment presented in the first 11 chapters of this book.

The first book included a small number of examples of application of risk assessment (legal, fate-transport, qualitative, time series, financial, plant construction, and others). Since its publication, there has been a clamor from numerous business, legal, and academic venues for a greater number of more detailed, real-life examples of the principles and techniques set out in that volume. The purpose of this book is to quell that din.

Risk and uncertainty examples outlined in this treatise are more comprehensive and meticulous than those delineated in the first book. Even so, risk assessment renditions in this volume still are generally elementary relative to most real-world applications of risk technology and processes. I had to strike a balance. The models described herein had to be detailed enough to have credibility and ring true with practitioners in the various fields represented by the examples. However, exceedingly long, complex, and minutely detailed risk models do not make stellar examples. Such complex applications generally are too "clumsy" to present, too lengthy, and numbingly esoteric. A point that is attempting to be made in such a model typically is lost because the logic thread is of too great a length and tied in far too great a number of cognitive knots. Therefore, the examples given here are just detailed and comprehensive enough to have credibility while hopefully retaining the quality of being eminently readable.

REALISM

I also have tried to inject an element of business-life realism into each narrative. Have you ever been put into the position of needing to accomplish a goal or deliver a product without having been afforded sufficient human and/or financial resources? Have you required crucial input from individuals who are not obligated to comply with your request for help? Have you ever been compelled to make a decision with far fewer facts than you think are required? Have you ever been given all the responsibility to make something happen but no authority to do so? Yes? Well, me too!

Each example is presented as a narrative. In most narratives, I follow a fictitious individual or set of individuals through the process of designing and enacting a risk model and interpreting the output. The narrative approach allows me to present serious subject matter in a light-hearted and readable manner. In addition, the narrative format facilitates the injection of real-world scenarios (such as those mentioned in the preceding paragraph) and to detail how such situations are handled in a business and risk assessment context. Although relatively simple, the presented risk models are real and the business and human-interaction discourses are true to life.

MODELS, VALIDATION, AND PRECISION

Most examples presented in this book include at least some computer code that encapsulates the logic of the risk model or models. The "language" in which these computer programs are presented is that of a risk assessment system devised by myself and my esteemed colleagues (Mike A. Long and Phil Hammond). Although the risk/uncertainty models presented here were generated using a software system not generally available, the models and the logic therein can be easily reproduced, for the most part, using any of a number of commercially available risk-related software systems.

The language in which the programs are presented is irrelevant. If presented in C++, undoubtedly someone would want them converted to FORTRAN (imagine that!). If presented in BASIC, a cadre of readers would prefer to see them described in JAVA. And so it goes.

The computer programs have been written using variable names that are explicit and understandable. The logic is laid out "from top to bottom," including comments in the code and a list of detailed variable definitions. This has been done to facilitate the translation of the model logic into any language. It is the logic, not the language, that is paramount. Admittedly, some user-written macros or subroutines might have to be composed to emulate certain system functions in the code presented (for example, the resampling of a distribution across time periods within a single Monte Carlo iteration).

A related point is that *there is nothing special about the risk models presented here*. The variables used and the codes displayed are but one set of variables and programs that might have been used to attack a given problem. I do not put forward these models as "the way to do it"; rather, these models are meant to instill in the reader that "it can be done." This is just one way, and not necessarily the best way, to accomplish the task.

Validation of the models can, in most instances, be done only over time. Attempting to validate the model relative to some other path that might have been taken or another model that might have been used is folly. You rarely, if ever, have the opportunity to know the result(s) of the "road not taken." In the case of many of the scenarios described in this book, the only validation possible is to deem whether the decisions made using model output were, in the end, good decisions for the company. If the majority of the courses taken result in economic prosperity for the company, then the models are vindicated on those grounds. You may argue that the company might have done just as well or better had they not even considered undertaking the risk-model route. You are welcome to do that.

Some of the model-output parameters are represented by numbers that have more "precision" than the numbers that went into creating them. This is a function of the software utilized to represent the output parameters. It allows the specification of X significant digits, but not how many are on either side of the decimal point. Therefore, to accommodate coefficients of relatively great magnitude, many significant figures had to be specified. This has the unfortunate and unintended effect for coefficients of relatively small magnitud, to have many more digits to the right of the decimal place than are warranted. The "significant figure" problem is not an easy one to dynamically adjust in the off-the-shelf graphics package employed here. Therefore, like so many other things in life, I decided not to like it, but just to live with it. The reader should know that I am aware of this problem but decided to chance the wrath of the readers rather than extend by a year the production of this volume. Your patience and understanding are requested and appreciated.

VALUE

As the reader, you will note that in each chapter there is an emphasis on the value of the opportunity that is the subject of the example. Value should be, but too often is not, the focus of a risk assessment model. Most models I have seen that are created by others tend to stop short of the calculation-of-value step. That is, if a risk model starts out as an environmental assessment, the answer (output) variables tend to be dosages of toxins, concentrations of an element in the environment, and so on. Likewise, construction models typically result in variables that express the cost and scheduling aspects of a project. Legal assessments tend to culminate in assessments of damages. Marketing models generate results that speak to the demographics, sales volumes, and margins related to a given scenario. The point is, businesses generally wish to make (and should make) tactical and strategic decisions based on the value of the opportunity to the corporation.

In this book I tend to focus on the net present value (NPV) of opportunities as the indicator of value. Measures of value such as internal rate of return (IRR), discounted return on investment (DROI), and return on capital employed (ROCE) can be equally valid in a given situation. The point is that nearly any risk assessment—whether it begins as a legal, technical, environmental, business, construction/manufacturing, or some other type—can and should generate a measure of value as an output parameter. This not only gives a corporation a true measure of the worth of an opportunity over time, but also affords a common basis upon which the corporation can compare, rank, and manage a portfolio of diverse entities. I am a firm believer in the philosophy of measuring the value of projects and opportunities, and the risk assessment examples in this book reflect that bent.

Section 1

Examples

1 Two Approaches to Solving Decision Trees — Class-Action Suit Example

CONTENTS

INTRODUCTION

Classical solution of a decision tree involves beginning at the "end" leaf nodes and multiplying leaf-node values by branch probabilities. The products of such multiplications are summed until we reach the "root" of the tree where the sum of the products results in an expected value for the tree.

Use of single-valued (i.e., deterministic) decision trees, especially for legal analyses and decisions, is a less-than-optimal approach for at least three reasons. Other concerns such as the inability of deterministic decision trees to consider the effects of "soft" issues through the use of chance of failure (see Chapter 16) also lend to the inadequacy in this arena. However, only three major drawbacks will be discussed briefly here.

The first foible relates to how a single erroneous assumption as a leaf-node value can render useless the calculated decision tree result. For example, most legal logic trains are lengthy. Any decision tree for a real-life legal analysis is really a "decision bush" with many (sometimes hundreds) of branches and associated leaf nodes. When using a deterministic tree, attorneys and clients must discuss and settle upon a single value at each leaf node. Because in the classical solution of the decision tree all leaf-node values and probabilities are multiplied together, a single incorrect value can invalidate the calculated result. In many instances, this error is difficult to detect and the problem goes unresolved. Use of ranges for values and accounting for dependence between parameters (see below) significantly alleviates this problem.

A second impediment to the use of deterministic decision trees is the inability to practically determine a reasonable range of results. That is, each solution of the decision tree results in a single calculated answer. In a tree that might contain hundreds of branches, it is an impractical task to change a single value at a given node and rerun the model to determine the impact of the change on the calculated answer. In addition, it often is necessary to change more than one leaf-node value for a given solution of the tree. It should be obvious that, in a tree that contains tens or hundreds of branches and nodes, the number of possible permutations can be in the thousands (or even millions, depending on the size of the tree). Attempting to generate a complete range of possible outcomes in this manner simply is not practical.

A third, but by no means final, glitch associated with the use of deterministic decision trees in legal analyses relates to the dependence between variables. In a deterministic tree, each value must be arrived at by employment of in-depth conversations. This mechanism is no different when distributions are used. When discussing the value for each node, humans must keep in mind the values they have entered for all other related nodes. For example, if we at some point in the decision tree have to enter the possible penalties associated with a judgment, we might have to relate that number to a previously entered leaf-node value (somewhere else in the tree) that indicates upon how many counts the client was deemed guilty. Typically, when approaching a decision tree in a deterministic manner, it is up to the humans, when changing values at leaf nodes, to keep in mind the relationship of those values to other values in the tree.

Using a probabilistic approach to solve decision trees does not absolve the tree makers from such considerations. However, most probabilistic software packages (those worth their salt, so to speak) contain a mechanism by which the tree builders can assign dependencies between variables. The tree can be instructed to select a value for a given leaf node that is reasonable with respect to other leaf-node values already selected. This means that consideration of dependencies needs to be addressed only once — when initially setting up the tree. Subsequent runs (or, in the case of Monte Carlo analysis, subsequent iterations) will honor the dependencies established. This is a tremendous benefit and time saver.

Solving decision trees probabilistically simply replaces the leaf-node deterministic values (and sometimes the branch probabilities, but this is a more intractable problem) with distributions, and the tree is solved many times. On each solution of the tree, a random grab is made from each leaf-node distribution, and the expected value is calculated in the usual way. Repeated random grabs and solutions of the tree result in a distribution of expected values.

The range of the expected values that results from this process is not representative of the full range of possibilities that one might encounter employing the decision tree in real life. An alternative method of probabilistic branching without using the branch probabilities as multipliers yields much more realistic results for a class of problems.

Decision trees are an often-used vehicle for analysis of legal strategies and cases. For example, an attorney might consider the following logic for a possible litigation:

> "We might settle the case. If we don't settle, we would take the case to court where we have a pretty fair chance of winning. However, if we lose the case, we will be subject to damages which could be of two kinds ...."

This type of logic often is diagramed as a decision tree. Even when probabilistic techniques are employed, solving such a decision tree in the classical manner results in a range of solution values that is not representative of real life. Neither the range of answers nor the associated probabilities are what the attorney really wants. In the following example I will demonstrate the difference between the classical approach and the probabilistic branching method for solving decision trees. A class-action-suit scenario will be used, but this logic holds true for decision trees that are applied to almost any problem.

BUILDING THE DECISION TREE

Perry is an attorney for the law firm of Lamb, Curry, and Rice. Regarding a major class action suit, Perry has been assigned the task of analyzing the situation and is to determine whether it is economically more desirable to settle the case or to take it to court. Perry has done his homework and has determined that the major aspects of the case are thus:

- We could settle the case for a large sum of cash.
- We could fight the case in court.
- If we go to court, we could win and our damages would be minimized.
- If we lose the case in court, we might be held responsible for physical damages only.
- If we lose the case, we might be held responsible for physical damages and, in addition, be made to pay punitive damages.
- If we are made to pay punitive damages, they may relate only to the contemporary physical damages.
- If, however, the jury decides that we are culpable for past damages, we might be made to pay punitive damages of a retroactive nature. Such punitive damages could be of much greater magnitude.

At first blush, this seems simple to Perry and he quickly generates the diagram shown in Figure 1.1. Now comes the more arduous and subjective task of populating the tree with probabilities and leaf-node consequences. To do this, Perry realizes that he will need the help of others. Perry convenes a meeting of his fellow attorneys.

At the meeting the group agrees on the branch probabilities and deterministic leaf-node values shown in Figure 1.2. Recalling from his college days the proper method for solving a decision tree, Perry quickly calculates the tree's expected value (EV) by applying the following calculation:

$$(0.3 \times 1.5) + 0.7 \times ((0.2 \times 3.5) + (0.8 \times ((0.7 \times 3.3) + (0.3 \times 6.5)))) = 3.326 \quad (1.1)$$

The only thing you can know for sure about the single-valued answer of a deterministic result (especially one carried out to three decimal places) is that it is not the value that real life will yield. Perry surmises that what he really wants is a probabilistic analysis of this case. To get such an analysis, Perry approaches Mason, the corporate risk expert. Mason advises Perry that he can easily convert the decision tree to a probabilistic model, but that Perry will have to replace most or all of the deterministic leaf-node values with ranges of values (i.e., distributions).

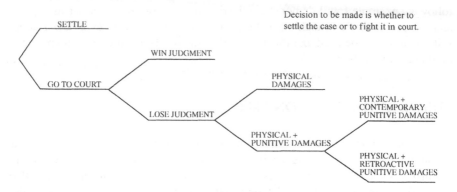

FIGURE 1.1 Basic decision tree for class-action suit.

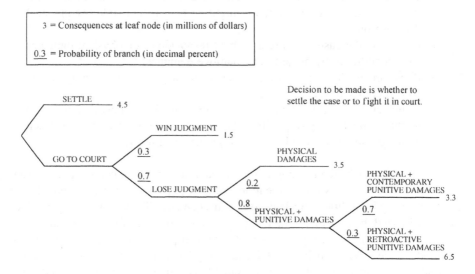

FIGURE 1.2 Class-action-suit decision tree with deterministic probabilities and leaf-node values.

After a week of meetings and data gathering, Perry returns to Mason with minimum, most likely, and maximum values for each leaf node in his decision tree. Mason will use the minimum, most likely, and maximum values in his proprietary software to build a distribution at each node. The resulting decision tree appears in Figure 1.3. The agreed-upon distributions are shown in Figures 1.4 through 1.8.

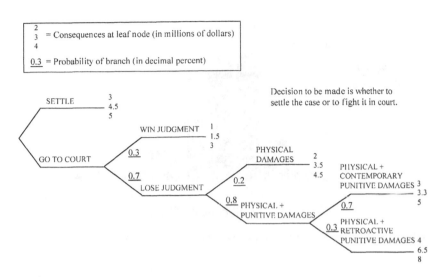

FIGURE 1.3 Class-action-suit decision tree with deterministic probabilities and leaf-node minimum, most likely, and maximum values for building distributions.

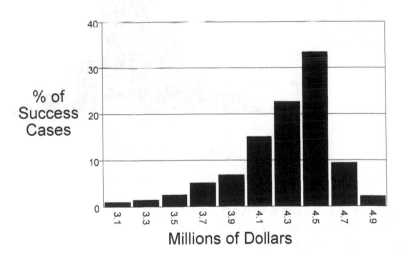

FIGURE 1.4 Distribution of settlement damages values.

The simple computer code generated by Mason for the probabilistic model is shown below.

```
PunDam = PhysPunProb * ((PhysConPunProb * PhysConPunDam) +
    (PhysRetPunProb * PhysRetPunDam));
```

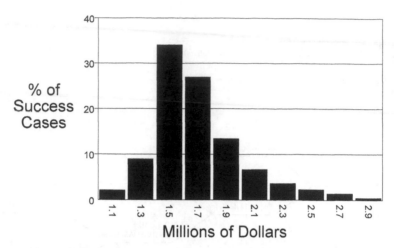

FIGURE 1.5 Leaf-node distribution for "win judgment" branch.

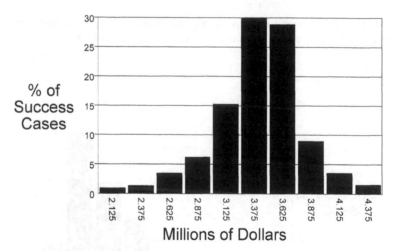

FIGURE 1.6 Leaf-node distribution for "physical damages" branch

PhysPunDam = LoseProb * (PunDam + (PhysProb * PhysDam));
Fight = (WinProb * WinDam) + PhysPunDam;
Delta = Fight − Settle;

where

    PunDam is the amount the firm will have to pay for punitive damages
    PhysPunProb is the probability that the firm will be held responsible for physical and punitive damages
    PhysConPunProb is the probability that the firm will be held responsible for payment of physical and contemporary punitive damages

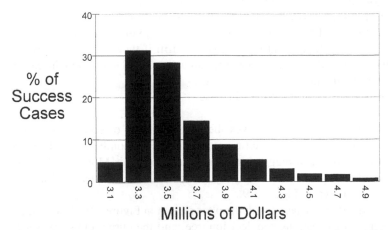

FIGURE 1.7 Leaf-node distribution for "physical and contemporary punitive damages" branch.

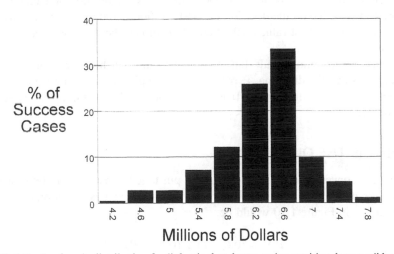

FIGURE 1.8 Leaf-node distribution for "physical and retroactive punitive damages" branch.

PhysConPunDam is the amount of physical and contemporary punitive damages the firm might have to pay

PhysRetPunProb is the probability that the firm will be held responsible for payment of physical and retroactive punitive damages

PhysRetPunDam is the amount of physical and retroactive punitive damages the firm might have to pay

PhysPunDam is the total amount of physical and punitive damages for which the firm might be held responsible

LoseProb is the probability of losing the judgment

PhysProb is the probability of having to pay only physical damages

PhysDam is the amount of physical damages alone
Fight is the total cost of fighting the case in court
WinProb is the probability that the firm will win the judgment
WinDam is the amount it will cost the firm if they win the judgment
Delta is the difference between the damages related to fighting the case relative
    to settlement
Settle is the cost of settling the case

This model iteratively solves the above system of equations a total of 1500 times. On each of the 1500 iterations, a value is randomly selected from each leaf-node distribution and plugged into the equations. This Monte Carlo process, then, yields 1500 answers. The frequency and cumulative frequency plots for the 1500 results are shown in Figure 1.9.

At first, Perry is happy with the plots shown in Figure 1.9. However, after staring at the input distributions, the decision tree, and the output plots, he begins to feel a bit uneasy.

Perry notes that the entire range of answers on, for example, the cumulative frequency plot is from about 2.8 to about 4.4. These values do not even come close to the lowest value on the decision tree ($1 million if Lamb, Curry, and Rice wins the case) or the highest value on the tree ($8 million if the firm has to pay retroactive punitive damages).

This is a troubling result, but Perry can't quite put his finger on the problem. To help resolve the dilemma, Perry creates a simple two-branch decision tree with deterministic (i.e., single-valued) leaf-node values. This simple tree is shown in Figure 1.10.

WHAT IS THE QUESTION?

Using the conventional decision-tree logic put together by Mason, the probabilities on the branches are used as multipliers for leaf-node values. Solution of the simple tree shown in Figure 1.10 would be:

$$E.V. = (0.6 \times 1) + (0.4 \times 10) = 4.6 \qquad (1.2)$$

In this equation, E.V. is the expected value for the tree which is 4.6. Perry notes that if the leaf-node values of 1 and 10 represented real-life consequences, a result of 4.6 is not a possibility. The real-life answers will either be 1 or 10. He also notes that this will be true regardless of the nonzero values assigned as probabilities for the decision-tree branches.

Perry thinks that the cause of the problem might lie in the fact that he is treating the problem in a deterministic manner. To test this, Perry asks Mason to build a new model for the simple decision tree shown in Figure 1.11 and to use the values minimum = 1, most likely = 2, maximum = 3 for the top-branch leaf-node distribution. For the bottom-branch leaf node, Perry gives Mason the distribution-building values of minimum = 8, most likely = 9, and maximum = 10. Mason builds and runs the model. Plots resulting from the model are shown in Figure 1.12.

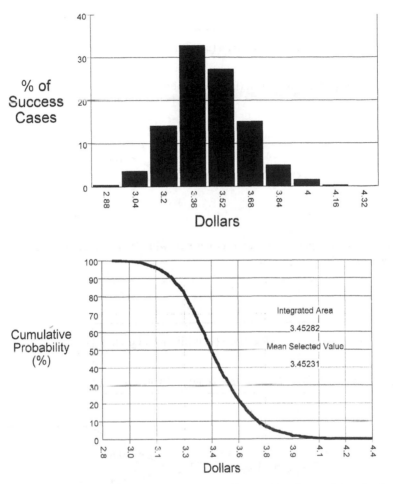

FIGURE 1.9 Frequency and cumulative frequency plots resulting from solving the decision tree using the Monte Carlo method and conventional decision-tree solving logic for damages for go to trial.

Inspection of the plots in Figure 1.12 quickly tells Perry that treating the problem probabilistically is also not the solution. Even though distributions were used for the two leaf nodes, the X-axis values in Figure 1.12 do not come close to the values used in the two distributions. In real life, taking the top branch of the decision tree (i.e., winning), regardless of the probability of taking that branch, a consequence between 1 and 3 would be realized. Likewise, regardless of the probability of taking the bottom branch, the real-life consequence of going down that path (i.e., losing) would be a value between 8 and 10. Conventional decision-tree logic that utilizes the probabilities as multipliers will not yield such results.

Perry now realizes that the conventional method for solving decision trees results in an expected value for the tree. Perry has to wrestle with the dilemma of just what question he really wants to have answered.

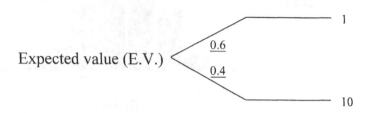

1 = Consequences at leaf node (in millions of dollars)

<u>0.6</u> = Probability of branch (in decimal percent)

Expected value (E.V.)

0.6

0.4

1

10

FIGURE 1.10 Simple two-branch decision tree with deterministic probabilities and leaf-node values.

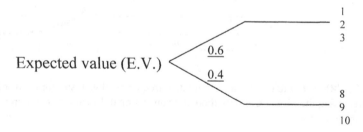

1	
2	= Consequences at leaf node (in millions of dollars)
3	

<u>0.6</u> = Probability of branch (in decimal percent)

Expected value (E.V.)

0.6

0.4

1
2
3

8
9
10

FIGURE 1.11 Simple two-branch decision tree with deterministic probabilities and distributions at leaf-node values.

After consulting with other attorneys and clients, Perry comes to the conclusion that the traditional method for solving decision trees — one that yields expected values — is not the answer to the actual question posed by a class-action-suit legal decision. All parties agree that for this application, the probabilities on the tree branches should not be used as multipliers for leaf-node values. Rather, a branch probability should be viewed as the chance that a branch will be taken. The decision to take a branch will be decided by, at each decision node, the generation of a random number. For example, in the decision tree shown in Figure 1.13, a random number between 0 and 1 would be generated at the decision point (point at which the branches

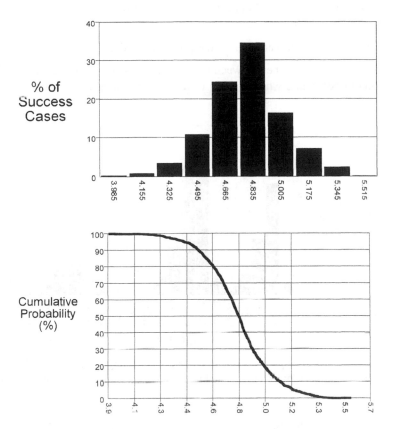

FIGURE 1.12 Frequency and cumulative frequency plots resulting from solving the simple two-branch decision tree for expected value using the Monte Carlo method and conventional decision-tree solving logic.

join — in this case, the E.V. position). If the random number generated is less than or equal to 0.6, then we would take the top branch and our consequence would be a number between 1 and 3. If the random number were greater than 0.6, then our consequence would be a value between 8 and 10. This process might be repeated hundreds or thousands of times in a Monte Carlo model resulting, in this case, in a bimodal distribution.

Since Perry now believes he knows what he wants, he approaches Mason with the concept and convinces him to write yet another program that captures the logic of probabilistic branching. Mason, too, is convinced that this is the right solution to the problem and produces the following program.

```
PunDam = if(rand() <= PhysRetPunProb, PhysRetPunDam,
   PhysConPunDam);
PhysPunDam = if(rand() <= PhysPunProb, PunDam, PhysDam);
```

Fight = if(rand() <= LoseProb, PhysPunDam, WinDam);
Delta = Fight − Settle;

where, in addition to the variables in the previous model,

rand is a random number between 0 and 1 that is generated at each decision
point to determine which branch of the decision tree that is taken.

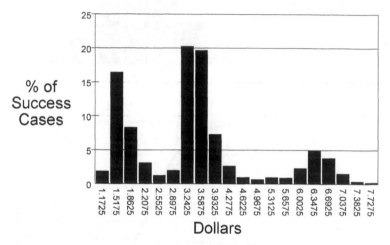

FIGURE 1.13 Frequency plot for go to trial damages resulting from using the Monte Carlo
method and probabilistic-branching method to solve the decision tree shown in Figure 1.3.

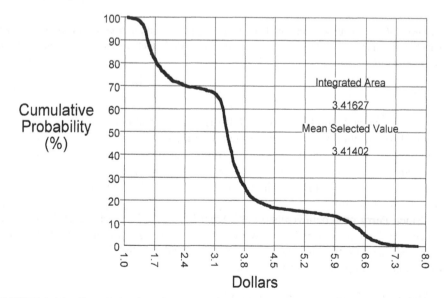

FIGURE 1.14 Cumulative frequency go to trial damages plot resulting from using the Monte
Carlo method and probabilistic-branching method to solve the decision tree shown in Figure 1.3.

INTERPRETATION OF THE PROBABILISTIC-
BRANCHING MODEL

Output plots from the probabilistic-branching model can be seen in Figures 1.13 and 1.14. After viewing and considering the results from this model, Perry is far more satisfied. It can be seen from Figures 1.13 and 1.14 that the extremes of the output distribution approach the extremes represented by the distributions of the leaf nodes. For example, the output-distribution values of greatest magnitude are near 10. This corresponds to the "physical and retroactive punitive damages" leaf-node-consequence distribution. It can be seen from the cumulative frequency plot that there is a very small probability of realizing a consequence near 10, but it is, nonetheless, a possibility. This small possibility corresponds to the relatively unlikely event that as decisions are made at each of the tree's decision nodes (moving from left to right), we will end up on this branch. The small likelihood of a value near 10 also reflects the relatively minute chance that even if we do end up on this branch, a random draw from the leaf-node distribution will result in selection of a value near 10 from that distribution. This emulates real life.

Cumulative frequency curves resulting from the classical method for solving decision trees tend to be smooth and "well behaved" (Figure 1.9). This is a natural consequence of calculating a "blended" expected value on each iterative solution of the decision tree. A cumulative frequency plot resulting from the probabilistic-branching method, however, is more likely to appear multi-modal. This results from the disparity in the magnitude of values at the various leaf notes (see Figure 1.3).

Figure 1.13 shows the frequency-plot equivalent of the cumulative frequency plot in Figure 1.14. Note that there are essentially three "modes," i.e., "peaks" in the distribution. The peak representing values of lowest magnitude is intermediate in height (frequency). This peak's X-axis position and relative frequency are mainly the results of the 30% chance of winning the judgment (see Figure 1.3). The peak of greatest frequency near the middle of the X-axis is of relatively great frequency because all but one branch of the decision tree has leaf-node values that could result in a value in the range represented by this peak. Similarly, the right-most peak in Figure 1.13 indicates that there is a relatively small chance that a value near the maximum of $8 million will be realized.

Perry is much more satisfied with his interpretation of the cumulative frequency curve resulting from the probabilistic-branching model. Interpretation of the curve indicates that there is a 100% chance of the case resulting in a cost of $1 million or more. The $1 million figure represents a real-life possibility if the firm wins the judgment. There exists about a 70% chance that damages from the case will be about $2 million or greater. There is about a 20% chance that damages will be $4 million or more. A small chance exists that a value near the maximum $8 million figure will be realized.

"Horizontal" sections of the cumulative frequency curve correspond to "low spots" in the frequency plot. These parts of the curve represent dollar ranges that are less likely to occur. Conversely, steeper "vertical" sections of the cumulative frequency curve correspond to "peaks" in the frequency display and indicate dollar ranges that are more likely to be realized. For Perry's purposes, which include communication with the clients, the results from the probabilistic-branching solution of the decision tree are much more satisfying than output from the classical decision-tree method.

SO, SO WHAT?

Technical advantages related to a probabilistic approach are discussed in the preceding paragraphs. Establishment of a realistic range of results and the association of probabilities with subranges within the total range of results are paramount. Other benefits also are realized.

One of the primary advantages of addressing legal problems with a probabilistic tool is the conversation that it mandates between the legal advisor and the client. Discussions regarding the establishment of single values to be used in a deterministic tree also must take place, but the quality and richness of the conversations are greatly enhanced when ranges are considered. Discussions of ranges forces the consideration of how "good" and how "bad" an outcome might be, and what it likely will be. In addition, much mutual understanding of the entire analysis process is fostered when ranges for multiple variables have to be considered.

Preempting surprise and dismay also is a benefit. When using deterministic tools, the answer is a single value (or a couple of hard-earned values). The client, then, expects (expected value) that the result of the legal process will be at or near the deterministic result. In real life, actual results can take on a wide range. When addressing the problem probabilistically, the process results in a complete range of outcomes. The attorney can discuss with the client the "good" outcomes and the "bad" outcomes, and all parties can understand just how those results could actually occur in real life (i.e., everyone understands the possible combination of events that could lead to extreme results). Clients understand that, although there may be a small chance of a "bad" outcome, it is still a possibility, and they realize just what set of circumstances could result in such a judgment.

The likelihood of results also helps the attorney and client prepare for the real-life outcome of a legal problem. In the example used in this chapter, Perry knows that there is a range of possible outcomes. However, within that range, there are subranges (i.e., places where the bars on the "answer-variable" histogram are "tall") that have a greater probability of harboring the actual result than do other subranges. This is important to know. The actions taken by a client to prepare for one sub-range of the output distribution might be very different from preparation for another subrange.

2 Terrorism Risk Models — Relative and Absolute Risk

CONTENTS

TERRORISM RELATIVE-RISK MODEL

Corporations and governments around the world realize that today's business and political environment is global. It is typical and necessary to do business in and with foreign countries that represent the entire spectrum of industrial, economic, and social development. Unfortunately, one of the concerns that must be addressed and understood when considering global entanglements is the level of threat represented by terrorist activities in the part of the world being considered.

Typically, concerns such as terrorism are measured by relative and absolute means. Governments and corporations might evaluate multiple opportunities in a number of countries. No country is free from terrorist threat, and no corporation would consider doing business with a purely terrorist state. Therefore, most countries being given serious consideration inherently represent some level of threat due to terrorism. A risk model that can relatively rank and compare serious contenders generally is a front-line tool employed to aid in the country-culling process. An example of such a model will be given here.

When the relative-ranking model has been applied and the short list of final contenders has been established, a second type of terrorism model can be used. As previously stated, every country represents some level of terrorist threat. It may be that some of the threats can be mitigated by taking appropriate preemptive measures. Typical of such measures are educating corporate travelers concerning customs of host countries, establishing a security office in a country, or constructing a corporate compound to house employees. Every mitigative measure comes at a price.

A risk model to calculate and integrate mitigative expenses can be generated. Because input to such absolute-cost models is more difficult to ascertain than the measures employed in the relative-risk model, the mitigation-cost model would likely be applied to only the most seriously considered countries. In the hypothetical example below, we will consider the scenario of a corporation attempting to assess the terrorist organizations in two countries in which they might do business. The processes applied and models built could equally be applied by government agencies.

WHAT IS THE QUESTION?

In its brainstorming session the corporate risk team was to define the question or questions to be addressed by the risk model. Several potential queries were posed.

1. What group of countries or regions of the world do we wish to avoid?
2. Of the many countries we are considering, which countries represent the lowest overall threat from terrorist activities?
3. What types of terrorist activities should we invest in countering regardless of the countries being considered?

Proponents of Question 1 argued that except for a few very large countries, market, political, economic, cultural, and business synergies typically dictate that the corporation invest in multiple countries in a region. If unstable governments, unfriendly religious factions, or other negative aspects exist in one or more countries in a region, then perhaps that portion of the potential market should be identified and avoided.

Those risk-team members who proposed Question 2 took a more parochial view. These individuals conceded that business usually is done on a regional basis, but that each country should be separately evaluated and ranked. Country-specific evaluations would avoid the problem of tainting an entire region by a single "bad apple" and allow the identification of potential business zones in noncontiguous countries.

Question 3 was put forth by team members who believed that all countries represent some degree of risk from terrorist activities. They argued that the primary terrorist activities (e.g., car bombs, kidnapping, or ambushing tourists) should be identified and a mitigation plan developed. They stated that if the primary threats could be successfully countered, many seemingly unfavorable countries might be classified as potential business sites. Therefore, they wanted to see the risk model used to identify the primary terrorist threats in various parts of the world and to estimate the cost of mitigating the threats.

After employing a hierarchy/voting process and following much debate, the team decided to initially address Question 2 of the risk model The corporate group charged with designing the contributing-factor diagram decided that the relative measure would be expressed on a scale of 1 to 10. Most input variables also would use the 1-to-10 scale.

BUILDING THE CONTRIBUTING-FACTOR DIAGRAM FOR THE RELATIVE-RANKING TERRORIST-THREAT RISK MODEL

Following much deliberation, the risk team concluded that, with regard to terrorism, corporate concerns are represented by three broad categories. These are

1. Organization
2. Funding
3. Experience and Technology Prowess

The structure and type of variables and associated information that comprise a contributing-factor diagram hinge significantly upon what type of model will be built from the diagram. Some members of the group, for example, advocated the use of a decision-tree model. Others, more mathematically inclined, proposed the use of a linear programming (LP) model. Still others indicated that a Monte Carlo model would be most appropriate.

It was decided that the decision-tree approach was not optimal. Even a probabilistic decision tree (as opposed to a deterministic tree) would not allow the flexibility required to relate the variables in the model. In addition, the deterministic decision tree software available to the group would result in a single expected value for the model, and they realized that management would require that output from the model be represented by ranges and associated probabilities so that they could better decide their course of action.

The LP approach allowed the flexibility of variable relationships but typically resulted in a deterministic result. In addition, the group considered that processes such as dependence, chance-of-failure, and risk-weighted values might be pertinent elements of the analysis. None among them felt qualified to program such required elements into a practical LP model.

Consensus was reached to use a Monte Carlo approach. This path would allow for the generation of ranges and probabilities, could easily accommodate dependence and chance of failure, and could automatically produce risk-weighted values that would be essential to the comparison of options. So, they proceeded to discuss the development of the model with a Monte Carlo approach in mind.

The Organization category is intended to include variables that relate to the degree to which terrorist groups in a country are organized, the size of the terrorist organization, and the logistical support offered the organization by the host government. The Funding category will include variables that express a terrorist organization's diversity of funding sources and the level of funding. The Experience/Technology category will be composed of variables dealing with the level of experience of organization members, the effectiveness of execution of terrorist activities by the organization, and the level of technical prowess exhibited by the group.

Three variables make up the Organization category with responses being expressed as a range of values (or a single value) between 1 and 10.

- Level of organization
- Size of organization
- Logistical support from host government

Adhering to the contributing-factor diagram-building rules and those for construction of a cogent risk model, the team next considered the task of generating definitions for each variable (question). Users of a risk model will take every opportunity to misinterpret anything the model builder says; *therefore, definitions for questions need to be comprehensive and succinct.*

The agreed-upon instruction for the variable Level of Organization is to enter a distribution of values on a 1-to-10 scale that represents the degree of organization of the terrorist group. A value of 1 indicates a loose coalition or highly fractionated and disorganized group. A value of 10 indicates a highly organized group exhibiting a tightly structured, hierarchical, and disciplined organization.

Risk-team members all agreed that knowing something about the level of organization of the terrorist group is important; however, there was some disagreement with regard to the meaning or impact of the 1 through 10 levels. A value of 1 indicates a relatively disorganized group. A 10 ranking portends a group that is highly structured and organized. Most risk-team members agreed that a highly organized group is the more dangerous and potent opponent. However, some team members argued that a highly fractionated group is one that is more difficult to extinguish or counter because each faction has to be independently addressed. In the end, the group agreed to instruct risk-model data suppliers to consider a value of 10 to be that which represents a group that is highly organized and presents maximum risk. Interpreters of model output agreed that they would interpret coefficients of maximum magnitude to represent maximum danger.

The direction for Size of Organization is to enter a distribution of values on a scale of 1 to 10 representing the size of the terrorist group. Values from 1 to 10 indicate

    1= a single individual
    2 = one to 5 members
    3 = up to 10 members
    4 = up to 20 members
    5 = up to 50 members
    6 = up to 75 members
    7 = up to 100 members
    8 = up to 150 members
    9 = up to 200 members
    10 = more than 200 members

Team members all conceded that the size of the organization is a primary risk parameter. Larger organizations, they agreed, are more difficult to counter and neutralize than are smaller organizations. A value of 10, therefore, represents a group whose membership is large and, accordingly, represents maximum threat. The size of the organization represented by each of the values from 1 to 10 was agreed upon by group members based upon their expertise with regard to the size of known terrorist networks throughout the world.

The instruction for Logistical Support from Host Government is to enter a distribution of values on a scale of 1 to 10 indicating the degree to which the host government is believed to supply the terrorist group with logistical support. Logistical support is defined as actions taken (or not taken) by the host government and includes refusing to extradite known terrorists, allowing terrorists to train at base camps within the host-country borders, allowing the terrorist organization to operate globally from their base camps, and other logistical aids. A value of 1 indicates no host-government support of the terrorist group. A value of 10 indicates complete complicity on the part of the host government.

One of the primary tactics in countering terrorist organizations is to bring to the attention of the host government and the world the fact that a terrorist group is operating within the borders of the host country. This is done in the hope of bringing international pressure upon the host government to expel the terrorists. Coercion such as this is effective, however, only when the host government is not in agreement with terrorist-group philosophy and activities. A value of 10 indicates a high degree of partnership between the terrorist organization and the host government and, therefore, represents a terrorist group that is going to be relatively difficult to neutralize.

The two variables that constitute the Funding category are

- Diversity of funding base and
- Level of funding

The variable concerning the Diversity of Funding Base requires one to enter a distribution (or single value) on a scale of 1 to 10 that represents your best estimate concerning the diversity of funding for the terrorist organization. A value of 1 indicates a single funding source. A value of 10 indicates a wide variety of sources including host and foreign governments, commercial sources, contributions from indigenous population, private benefactors, and other sources.

A fundamental maneuver in blunting terrorist activities is to eliminate sources of funding for terrorist organizations. Publishing lists of financial contributors, seizing or freezing bank accounts and other assets, and causing terrorist groups to consume financial resources on nonterrorist activities (such as repeatedly moving base camps) are a few popular ploys. Regardless of the level of funding, organizations that are single-sourced or that enjoy relatively few funding sources are more easily strangled financially than even anemically funded groups who enjoy broad-based financial support. Therefore, a measure of funding diversity is a critical element in the risk model.

The variable Level of Funding directs one to enter a distribution (or single value) on a scale of 1 to 10 that represents the best estimate concerning the level of funding available to the terrorist organization. A value of 1 indicates a funding level that is generally insufficient to support continuous operations by the terrorist organization.

A value of 10 indicates that funding is far in excess of the needs of the organization or that obtaining required funds presents no problem for the organization.

Even a relatively inept terrorist group can be annoyingly persistent or horrifyingly deadly if well funded. Political and religious influence, sophisticated weapons, and mercenaries are just a few of the favorite shopping-list items of the terrorist group with sufficient monetary resources. Independent of the number of funding sources, a primary measure of a terrorist group's ability to cause mayhem is the level of funding the group enjoys. Therefore, a funding-level variable is a primary element of a terrorist-threat risk model.

The third category in the terrorist risk model is Experience and Technological Prowess. Three variables make up this category.

- Level of experience of members
- Effectiveness of execution
- Level of technological prowess

Level of Experience of Members instructs one to enter a distribution of values (or a single value) on a scale of 1 to 10 indicating the level of experience in terrorist activities exhibited by terrorist-organization members. A value of 1 indicates that members are novices. An intermediate value of 5 indicates either an overall experience level between novice and expert or the existence of a relatively small number of experts among a bulk of novice members. A value of 10 signifies that the bulk of the members are experienced terrorists.

Experience in terrorist activities is a primary factor in determining the effectiveness, boldness, and political acumen of a terrorist organization. Experienced members make fewer mistakes, better plan their activities, better cover their tracks, and coordinate their activities with sympathetic entities more efficiently than do less experienced individuals or groups. A measure of this experience level will serve as a critical parameter in the risk model.

Risk-model parameter Effectiveness of Execution directs one to enter a distribution of values (or a single value) on a scale of 1 to 10 that indicates an estimate of how effectively the terrorist organization carries out its activities. A value of 1 indicates either an organization that poorly plans its terrorist attacks or one that is relatively inept in carrying out activities. A value of 10 indicates an organization that efficiently plans and executes its plots.

Independent of the level of experience of its members, a terrorist organization's efficiency in execution of its activities is a prime concern. There exist some poorly funded and relatively inexperienced organizations which, because of efficient planning, networking, indigenous-population support, and other characteristics, are highly efficient in their terrorist activities. At the opposite end of the spectrum there exist well-heeled groups that exhibit overt and clumsy execution styles. An independent estimate of the efficiency of execution for a given terrorist organization is a useful measure in the overall estimate of the threat represented by that particular organization.

The variable Level of Technical Prowess requests one to enter a distribution of values (or a single value) on a scale of 1 to 10 that represents an estimate of the technical capability displayed by the terrorist organization. A value of 1 indicates an organization that utilizes relatively crude and mundane technology to carry out attacks. Such crude techniques might include pipe bombs, car bombs, small-arms attacks, and suicide bombings. A value of 10 identifies an organization that utilizes state-of-the-art technology. Such organizations may resort to computer hacking, smart weapons, fire-and-forget munitions, satellite imagery and positioning, and other sophisticated techniques.

Independent of funding and experience levels, some terrorist groups tend toward high-tech tactics. In fact, some of the most poorly funded groups are those that tend toward high-technology antics. The cost of a computer and modem is low. However, the mayhem that can be caused by relatively inexperienced, poorly funded, but technically astute terrorist individuals or groups is very significant. Favorite targets of low-budget hackers are bank accounts, power grids, communication networks, and other computer-controlled systems and infrastructures.

Other organizations employ individuals who are relatively unskilled and inexperienced, but who utilize high-tech weapons that compensate for their relative dearth of personal sophistication. Terrorists lacking any formal education can quickly be taught to launch highly effective fire-and-forget projectiles or heat-seeking or radar-guided missiles. Therefore, a measure of technical prowess is a fundamental parameter in the risk model.

All variables and categories are included in the contributing-factor diagram shown in Figure 2.1. This diagram served as the blueprint for the risk model.

CATEGORY WEIGHTS

The risk model contains three separate categories: organization, funding, and experience/technology. Weights are used in the model to afford those applying the risk

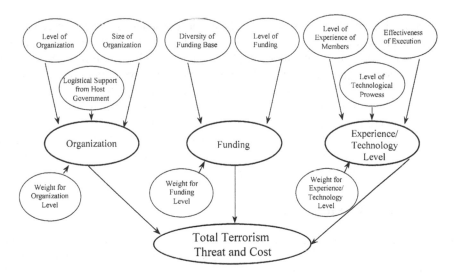

FIGURE 2.1 Contributing-factor diagram that serves as the basis for relative- and absolute-risk models.

model the ability to emphasize or de-emphasize the various model components. For example, if the terrorist group being evaluated is a longstanding and established organization known to have infiltrated government agencies and businesses, then the funding aspect may be moot. However, because we may have top-secret military installations in the country where the terrorist organization is known to operate, we are concerned about its ability to employ high-tech devices. Such devices could interrupt military communications, eavesdrop on such communications, or cause an electromagnetic pulse (EMP) that would damage sensitive and critical electrical components at the military compounds. Therefore, when evaluating this terrorist group, we would apply relatively little weight to the funding category and would weight more heavily the category representing the group's technical prowess.

Risk-team consensus on the units for category weights was that the weights should be expressed as a percentage represented by a value between 0 and 1 and the sum of the three weights must be 1. In addition to deciding on units, the team members realized that they must also come to agreement regarding the definition of the term *weight*. For example, just a few of the reasons that a category might be assigned a weight near 1 might be

1. The category represents certain parameters for which the terrorist group is well known.
2. The category includes variables that, for the terrorist group in question, will be particularly difficult to counter.
3. The category represents concerns that the corporation should spend time and money attempting to mitigate.
4. The category includes risk factors that the corporation is ill equipped to counter and, therefore, represents problems about which the corporation could do little.

Following much discussion and debate, the risk-team members agreed that a category should receive a relatively low weight when that category represents parameters about which the corporation could do little or nothing to mitigate the risks. A relatively low weight should be assigned to such a category regardless of the magnitude of the coefficients representing individual parameters that make up the catagory. Conversely, a category should be assigned a relatively great weight when the category is composed of parameters that represent risks the corporation has a reasonable chance of mitigating if sufficient resources are applied. The weights should be assigned to a category regardless of the magnitude of the coefficients representing individual parameters that make up the category.

RELATIVE-RISK MODEL EQUATIONS

Risk-team members realized that the risk model they were engineering likely would find application in all corporate offices around the world. Local corporate branches had for years been clamoring for some reasonable, consistent, and user-friendly means of assessing terrorist-activity risk, at least for budgetary purposes. Team members were cognizant of the fact that their model must be simultaneously comprehensive and simple. Stressing these simple guidelines, the team decided to integrate all distributions and weights using the following elementary set of equations.

org_risk = ((level_org + size_org + log_support)/3) * org_weight;
fund_risk = ((diverse_fund + level_fund)/2) * fund_weight;
exp_tech_risk = ((level_exp + effect_exec + tech_level)/3) * exp_tech_weight;
total_risk = (org_risk + fund_risk + exp_tech_risk);

where

org_risk is the organizational risk
level_org is the level of organization
size_org is the size of the organization
log_support is the logistical support supplied by the host government
org_weight is the weight applied to the organization category
fund_risk is the risk associated with funding
diverse_funding is the diversity of funding
level_fund is the level of funding
fund_weight is the weight applied to the funding category
exp_tech_risk is the risk associated with experience and technical prowess
level_exp is the level of experience
effect_exec is the effectiveness of execution
tech_level is the level of technical prowess
exp_tech_weight is the weight applied to the experience/technical prowess
    category
total_risk is the total calculated relative risk

RELATIVE-RISK MODEL APPLIED TO TERRORIST ORGANIZATION #1

First to be considered and evaluated by the risk model was a terrorist group that over the years has become ingrained in the social fabric of the host country. Many of the major public services, banks, and manufacturing businesses are either owned by members of the organization or "protected" by the group. Local governments also are known to be controlled or heavily influenced by the terrorist group. The national government has mounted earnest but token and inconsistent efforts to eliminate or control the group. All efforts have been in vain.

The corporation was considering establishing a branch administrative office, manufacturing facility, and distribution/sales center in the host country. Corporate policy and integrity forbid the payment of bribes or "protection" money. The terrorist organization has been known to resort to extortion, kidnapping, high- and low-tech attacks on physical assets, and murder to force compliance and payment from corporations. Some of the facilities the corporation might establish in the host country would be especially susceptible and vulnerable to terrorist activities and attacks.

In a meeting of the corporate risk team, each of the parameters for the relative-risk model was discussed. A range of distribution values and a peakedness value were agreed upon.

The first parameter relates to the degree to which the terrorist group is organized. Although the terrorist group is geographically scattered and multifaceted with many branches and subgroups, there exist both written and unwritten codes by which all members conduct themselves. All branches ultimately answer to a central authority or "boss," and misconduct within the organization is not well tolerated. Therefore, everyone on the risk team agreed that the group should be characterized as highly organized. Some risk-team members argued, however, that because the organization is typically run at the local level with individual local bosses having little or nothing to do with one another, the level of organization should be characterized as high, but not the maximum possible. The group decided on the following distribution-building values.

Minimum = 4
Most likely = 8
Maximum = 9

Peakedness is expressed as a value from 0 to 10 and is an indication of confidence in the most likely value. A value of 10 stresses the most likely value in the distribution, making it and values near it more likely to be selected in the Monte Carlo process. A peakedness value of 0 significantly de-emphasizes the most likely value. See Chapter 15 for a more comprehensive explanation. A peakedness value of 10 was established to emphasize the value of 8. The resulting distribution is depicted in Figure 2.2.

Size of the organization was next discussed. Risk-team members initially were tempted to characterize the size of the terrorist group with a value of 10. However, some team members argued that in some instances the corporation will likely be dueling with only a specific faction within the organization. They indicated that local "cells" of the terrorist group typically are composed of fewer than 100 members. Considering this, the risk team agreed to represent the size parameter with a truncated distribution built from the values

Minimum = 7
Most likely = 10

Peakedness was set to 5. Relatively inflating the 7 value using a peakedness of 5 placated those who argued that local cells of the terrorist organization are more significant. This distribution is shown in Figure 2.3.

Host-government logistical support was next discussed. In the considered country, the government occasionally attempts to round up terrorist-group members for prosecution. The government gives no overt aid to the organization. However, the government is forced to turn a blind eye toward much of the terrorist group's activities, because many of the necessary societal and business functions in the country are owned or controlled by organization members. For example, a severe crackdown by the military likely would cause the collapse of the banking, transportation, and other necessary industries. Therefore, although the host government does nothing to logistically aid the terrorist group, it does relatively little to attempt to eradicate the organization.

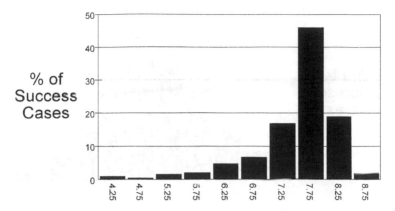

FIGURE 2.2 Distribution for level of organization for terrorist group #1.

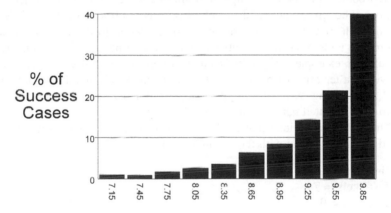

FIGURE 2.3 Distribution for size of organization for terrorist group #1.

Risk-team members agreed that midrange values should be used to characterize this parameter and agreed on the following values.

Minimum = 3
Most likely = 5
Maximum = 7

Peakedness was set to a value of 2. This reflected the group's feeling that sometimes the host government does in fact attempt to counter the group but at other times does nothing to punish group members for overt terrorist actions. The resulting distribution is shown in Figure 2.4.

The second risk-model category addresses funding considerations. First to be evaluated was the breadth of the funding base that supports terrorist-organization activities. Because group members are integrated into society as well as into most types of businesses and local governments, the funding base is very broad within the host country. However, funding for the organization seems to stop at the host-

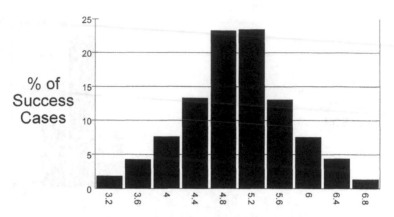

FIGURE 2.4 Distribution for host-government logistical support for terrorist group #1.

country border. International support traditionally has not been forthcoming. In addition, although the potential monetary base is multifaceted, it has, in the past, been difficult to convince one arm of the organization to tap its limited financial resources to underwrite the activities of an unrelated part of the organization. It is difficult, for example, to get the bankers of the organization to get financially involved in a terrorist act perpetrated by the trucking industry. Therefore, although broad, the funding base for any given organizational faction is somewhat limited due to the disjointed nature of the "siloed" organization.

The risk team decided that the breadth-of-funding parameter should be represented by a distribution with the following values.

Minimum = 5
Most likely = 7
Maximum = 9

A peakedness value of 1 was selected because a low peakedness number, in the judgment of the risk team, appropriately inflates the tails of the distribution. Inflation of the lower end of the distribution (values near 5) indicates the fractionated nature of the group. Emphasizing values near 9 with the peakedness value of 1 reflected the team's feeling that if matters were serious enough to attract the attention of the overall organization's "boss," significant resources could be gleaned from a wide array of sources. The resulting distribution is shown in Figure 2.5.

Level of funding was next considered. As explained previously, the organization is strongly segmented. Each faction generally draws funds only from its local base of support, that is, "protection" money from local businesses, "user fees" for essential services, etc. The host country is not a wealthy one. Populations in provincial towns and villages, therefore, have little to give. As a consequence, local organization members tend to live not much better than anyone else, the real wealth of the organization being concentrated near the top of the hierarchy. Therefore, funding levels tend to be rather meager unless the "bosses" at the head of the organization deem it necessary to sponsor an infusion of funds for one of the local groups.

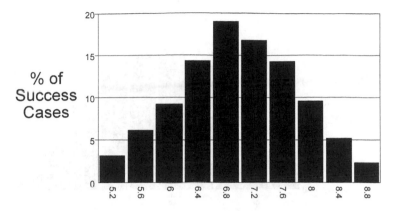

FIGURE 2.5 Distribution for diversity of funding base for terrorist group #1.

The risk team decided that the distribution for the level-of-funding variable should mimic that for the breadth of the funding base. The following values were used to build the distribution.

Minimum = 5
Most likely = 7
Maximum = 9

For the same reasons as the funding-base parameter, a peakedness value of 1 was deemed appropriate. That is, inflation of the values near 5 indicates that funding generally is adequate. The most likely value of 7 signifies that local groups can occasionally squeeze more funds from local businesses when needed or can attract some funding from another local faction if necessary. Inflation of the values near 9 indicates that if serious trouble is encountered, local arms of the organization can count on the backing of the parent organization. The distribution depicting funding level is shown in Figure 2.6.

A final category to be considered in the relative-risk model is that concerned with the experience level of organization members and the general technical prowess exhibited by the group overall. Experience in planning and executing terrorist activities is considered first in this category.

As mentioned previously, organization members also are, for the most part, prominent members of local governments, business, and society in general. Members, therefore, tend to grow up within the organization, getting their first experiences in terrorist activities at an early age. Being reared within the organization produces adult members with many years of experience to draw upon. More significant, however, is the mindset exhibited by members who have grown up in the organization. Because terrorist activities of all sorts have been part of their daily experiences throughout their lives, organization members accept terrorist behavior as a way of life. This is significant because it produces a terrorist who lacks the sense that such activities are morally and socially abhorrent. Members without such moral and social guideposts tend to resort to repugnant acts as a matter of course and do not yield

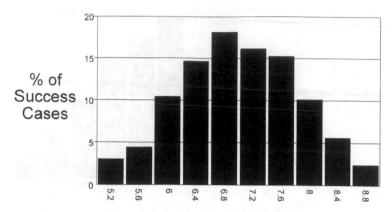

FIGURE 2.6 Distribution for level of funding for terrorist group #1.

to moral arguments. Such individuals are among the most dangerous and difficult-to-negotiate-with terrorists in the world. The organization, however, as part of a tutelage system, tends to assign the execution of many of its campaigns to younger members so that they might gain experience.

After considering and discussing these experience-related attributes, corporate risk-team members unanimously decided that this terrorist organization should be represented as having highly experienced members who direct less experienced individuals in the execution of terrorist acts. Team members decided that a truncated distribution should be used to represent organizational experience. The following distribution-building values were agreed upon.

Minimum = 5
Most likely = 10

A peakedness value of 10 was established so that the distribution value of 10 would be emphasized. The resulting distribution is shown in Figure 2.7.

Effectiveness of execution is next considered in the relative-risk model. Effectiveness is considered separately from experience because some terrorist organizations harbor highly experienced members who plan operations but do not personally take part in the execution of the terrorist activity. Such organizations, while incorporating experienced members, exhibit a wide range of effectiveness of execution when the actual terrorist act is left to be executed by less experienced people.

The corporate risk team knew that the terrorist organization being considered is strongly hierarchical. Senior members who have "come up through the ranks" are experienced, typically plan terrorist strikes, but are rarely involved in the execution phase. Enactment of the terrorist activity is reserved for "up and coming" members who desire to prove their merit by carrying out the plots. When a junior member has successfully executed one or two terrorist acts, that member expects to move up in rank and be replaced by a younger and less experienced person. So, while it is true that the organization harbors very experienced individuals, attacks are generally carried out by less experienced junior members. Therefore, the success-of-

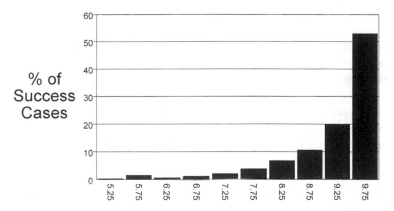

FIGURE 2.7 Distribution for level of member experience for terrorist group #1.

execution record for the organization over the past decade is rather spotty. In fact, if the host government would take a serious attitude toward bringing a halt to the organization's activities, the organization's success record with regard to executing its terrorist plots could be greatly impacted.

Corporate risk-team members agreed that on a scale of 1 to 10, the organization should be represented by "middle-of-the-road" values. They settled on the following distribution-building parameters.

Minimum = 3.5
Most likely = 5.0
Maximum = 6.5

A peakedness value of 1 was assigned because the group wished to emphasize the relatively great probability of values in the tails of the distributions. The distribution is shown in Figure 2.8.

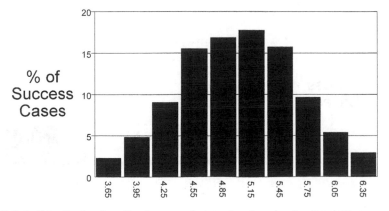

FIGURE 2.8 Distribution for effectiveness of execution for terrorist group #1.

Level of technical prowess is the last parameter to be considered in the relative-risk model. Some terrorist organizations, while not highly experienced nor well funded, carry out very sophisticated terrorist acts. These include computer hacking into critical databases, the use of "smart" and fire-and-forget munitions, and elaborate kidnapping and extortion activities utilizing sophisticated communication and other electronic gear.

Although the terrorist organization being assessed has in the past employed outside-the-organization "hired hands" when technical sophistication was required, the organization historically has employed relatively low-tech tactics. Beatings with bats, threats and attacks with handguns, use of pipe bombs, and other relatively crude tactics constitute their repertoire of methodologies for intimidation. However, they have not been reticent to "buy" high-tech help when deemed necessary.

Risk-team members decided that the distribution to represent the technical prowess of the organization should be bimodal. The values for the two-mode distribution were agreed to be

Minimum = 2.5
Most likely = 3.5
Maximum = 5

Minimum = 6
Most likely = 7.5
Maximum = 8.0

Peakedness values of 3 for the first distribution and 7 for the second distribution were established. Relative weights assigned to the two parts of the bimodal distribution were agreed to be 80 and 20. These values, it was decided, represented the percentage of times the organization instituted relatively low-tech operations vs. hiring high-tech help. The resulting bimodal distribution is shown in Figure 2.9.

Finally, for the model, the risk team needed to decide the relative weights that would be assigned to each category. As delineated previously, members of the

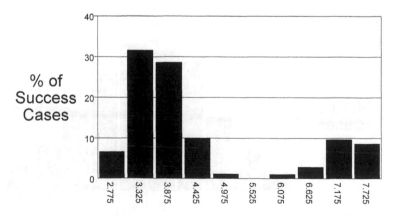

FIGURE 2.9 Distribution for level of technical prowess for terrorist group #1.

terrorist group in question are part of the social, political, and business fabric of the nation. The terrorist group provides, in many communities, banking, transportation, refuse disposal, and other essential services. Therefore, any attempt to disrupt the organizational facet of the terrorist group is an attempt to degrade the functioning and efficiency of communities. It was deemed by the risk team that it would be folly to think that a single corporation could have a significant negative impact on the organizational aspects of such a group.

Likewise, funding for the terrorist group is gleaned from a wide variety of political and business sources and through various intimidation tactics. While there is little doubt that the corporation could put in place policies and practices that might somewhat negatively impact funding for the group, the risk team decided that no significant diminution of funding could be brought about by corporate actions.

Contrary to the decisions made concerning the organizational and funding aspects, the risk team concluded that the corporation might, in fact, be capable of putting a dent in the terrorist organization's use of technology against the corporation. Although the host government is reticent to enact overt measures to counter the terrorist organization, it is willing to work "behind the scenes" to fight terrorism. Because the corporation will be a significant contributor to the host-country tax base, corporate executives are confident they can convince government officials to begin, at ports of entry, an import-screening program using high-tech equipment capable of detecting plastic explosives, biological agents, and other often-used terrorist materials. Such screening procedures would be put in place at all corporate locations.

In addition, the corporation plans to erect, around each corporate facility, low-tech decorative concrete barriers. Barriers will be positioned to hinder car-bomb attacks. To guard against kidnapping attempts, the corporate headquarters site will include a first-class corporate compound at which resident and visiting corporate executives and their families will live. Armored cars will be provided for travel outside the compound. Guards at the corporate sites will be brought from the United States in an attempt to minimize the influence of the terrorist organization on corporate security. These and a host of other actions will be taken by the corporation to guard against the use of high-tech and other weapons in the terrorist arsenal.

The risk team agreed that higher weights should be applied to those categories that could be significantly affected by application of corporate resources. Team members decided to give a weight of 0.15 to the organization category, 0.15 to the funding category, and 0.7 to the experience/technological prowess category. A summary of model input parameters appears in Table 2.1.

RELATIVE-RISK MODEL RESULTS FROM EVALUATION OF TERRORIST ORGANIZATION #1

Risk-team members input to the risk model data generated for the first terrorist organization (designated terrorist organization #1). The resulting output distributions can be seen in Figures 2.10 through 2.13. Figure 2.13 shows that the modal risk value for this terrorist organization is around 6.5, as is the risk-weighted value. The distribution has a minimum value of about 4.7 and a maximum value of about 7.8. Risk-team members noted that variables in the organization and funding categories

TABLE 2.1
Input Parameters for Relative Risk Model — Terrorist Organization #1

Category/Parameter	Minimum	Most Likely	Maximum	Peakedness
Organization Category				
Level of Organization	4	8	9	10
Size of Organization	7	10		5
Logistical Support from Host Government	3	5	7	2
Funding Category				
Diversity of Funding Base	5	7	9	1
Level of Funding	5	7	9	1
Experience/Technology Category				
Level of Experience of Members	5	10		10
Effectiveness of Execution	3.5	5	6.5	1
Level of Technological Prowess	2.5	3.5	5	3
	6	7.5		
Category Weights				
Organization Category Weight		0.15		
Funding Category Weight		0.15		
Experience/Technology Weight		0.7		

are represented by coefficients of relatively great magnitude, yet the weighted values for both categories are relatively paltry (see Figures 2.10 and 2.11). This, of course, is a reflection of the small-magnitude weights assigned to these categories. All plots and data generated will be used to compare this terrorist organization with the one to be considered next.

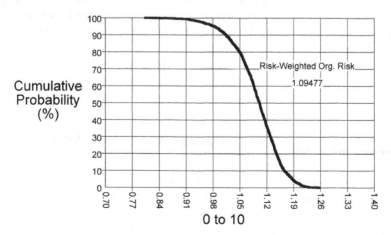

FIGURE 2.10 Cumulative frequency curve showing range of weighted risk associated with organizational aspects of terrorist group #1.

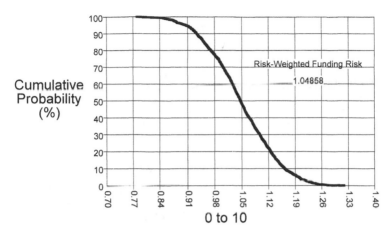

FIGURE 2.11 Cumulative frequency curve showing range of weighted risk associated with funding aspects of terrorist group #1.

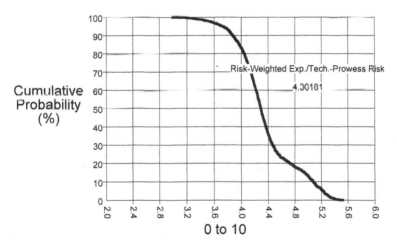

FIGURE 2.12 Cumulative frequency curve showing range of weighted risk associated with experience/technological prowess aspects of terrorist group #1.

RELATIVE-RISK MODEL APPLIED TO TERRORIST ORGANIZATION #2

Next to be considered was a terrorist organization that is headquartered in a host country. Like the previous situation, the corporation was considering establishment of a branch office in the host country, including administrative, manufacturing, and distribution functions.

Most aspects of this situation were, however, markedly different from the previously considered scenario. The host government is a religious-based, quasi-fascist, militaristic regime. The government holds onto power by holding elections in which

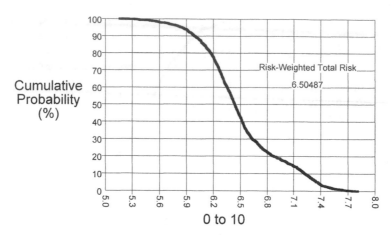

FIGURE 2.13 Cumulative frequency curve showing range of total weighted risk associated with terrorist group #1.

fear and intimidation dictate the outcome of the "voting." Most civil laws are drawn from religious doctrine.

The country was created a decade ago when a large coalition government collapsed, creating numerous independent states. Although the government in question has been rather repressive over the past decade, it has seen several of its sister states prosper by adopting a more moderate and externally focused posture. The country has, within the past year, approached several large corporations in an attempt to establish a fledgling manufacturing and general business base for its economy — hence the risk assessment.

Corporate planners believed that there existed considerable market potential in the area and seemed anxious to be first to initiate relations with the government and to establish a physical presence in the country. A significant stumbling block for this path, however, was the government's penchant for hosting and logistically supporting terrorist activities.

The terrorist organization in question is dominated by a single leader. Under the guise of a religious war, the organization head has railed against Western influences and has vowed the destruction of same. The terrorist organization itself is a coalition of independently operating cells. Religion and hatred of things Western serves as the binding force. In the past, the host government has provided the organization leader with financial and logistical support and has allowed the leader to establish base camps where organization members are indoctrinated and trained.

In keeping with the relative-risk model, the first attributes to be considered were organizational. Although the head of the terrorist organization pontificates for the group, he exercises no real power over the individual cells that compose the greater organization. Any control he enjoys is gleaned from his deteriorating relationship with the host government, a shared hatred of Western entities, and a religious view shared by him and members of the cells. Because of the leader's loose ties with member cells and because individual cells rarely collaborate and, many times, mistrust one another, risk-team members deemed the level of organization to be low. However, some risk-team members pointed out that because the organization is religiously based, a "call to arms" predicated on religious principles would be a powerful catalyst to unify the many cells. Therefore, an unlikely but high degree of organization cannot be discounted. They agreed that the level-of-organization variable should be represented by a distribution built from the following parameters.

Minimum = 1
Most likely = 2.5
Maximum = 9

It was agreed that a peakedness of 3 faithfully represented the relative frequencies of the most likely value and those in the tails of the distribution. The resulting distribution is shown in Figure 2.14.

It was difficult for risk-team members to estimate the organization's size because the terrorist group is an aggregate of loosely affiliated cells. In addition, individual cells tend to exist only when solicited to take action; i.e., there is no "standing army." When cells do congeal, they tend to have individual memberships no greater than about ten in number. International intelligence networks confirm that never have more than 2 to 3 cells been active at a given time; however, as previously noted, a religious "call to arms" could cause great numbers of cells to cooperate. Therefore, corporate risk-team members agreed that the corporation would most likely at any one time be dealing with anywhere from 1 to 50 terrorists, but in rare instances could have to deal with hundreds of members. They agreed that the organization-size distribution should be built from the following parameters.

Minimum = 1
Most likely = 3
Maximum = 9

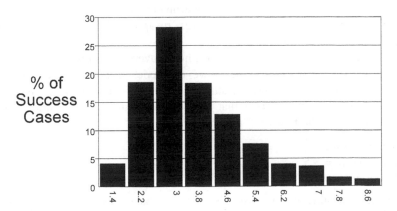

FIGURE 2.14 Distribution for level of organization for terrorist group #2.

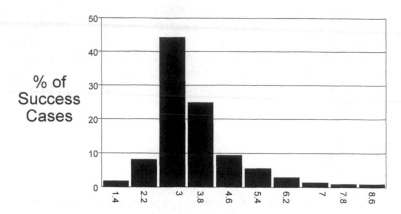

FIGURE 2.15 Distribution for size of organization for terrorist group #2.

A peakedness value of 8 was established, which emphasizes the midrange of the distribution but still allows reasonable probability for values near the distribution's extremes. The resulting distribution is shown in Figure 2.15.

In the past, host-government-supplied logistical support for the organization was strong. Shared belief systems and common political postures and agendas prompted government officials to overtly supply the terrorist group with real estate, money, munitions, and a blind eye when such munitions were employed against political entities, individuals, and organizations that were considered hostile to their aims. In recent years the proactive-terrorist position has resulted in the host government being characterized as a repressive and recalcitrant regime surrounded by political entities of a much more progressive and cooperative posture. Businesses and other governments, therefore, have tended to eschew investing in the host country.

Money may be the root of all evil, but it also is the basis for economic and social prosperity. Recent elections have deposed many hard-line host-government cabinet members, and signals of moderation recently have been given to ambassadors from the host-country president. Although the host government has not yet ousted terrorist-organization members, overt government support for such organizations in recent years has waned and, in some cases, disappeared. It is suspected, however, that clandestine support still exists.

Risk-team members agreed that overt host-government support of terrorist activities of all types is on the decline; however, moderate covert augmentation of the terrorist organization in question still could exist. After much discussion of the subject, team members agreed that the logistical-support distribution should be constructed from the following parameters.

Minimum = 2
Most likely = 3.5
Maximum = 8

A peakedness value of 4 was established to emphasize the values in the tails of the distribution relative to the most likely value. The resulting distribution is shown in Figure 2.16.

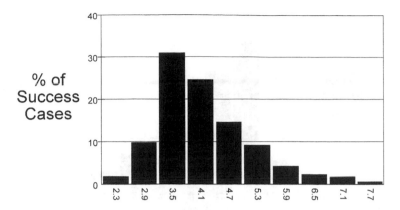

FIGURE 2.16 Distribution for host-government logistical support for terrorist group #2.

Funding was the next category to be considered in the relative-risk model, with diversity of funding first being addressed. The leader of the organization is known to collaborate with kith and kin in other countries; however, there has never been any evidence that terrorist-organization counterparts in other countries ever contributed to the coffers of the organization in question. In addition, the apparent sole funding stream for the organization has sprung from the organization leader's cozy relationship with host-government cabinet members. Such funding sources are dissipating, and the overt direction of cash or material from the government to the organization has in recent years become a political death knell for many government officials. However, if the host government found it necessary to carry out a terrorist activity to further its ends, it might bring pressure to bear on some sympathetic neighboring countries and other terrorist organizations to provide funds for operations.

Risk-team members, therefore, concluded that the diversity of the funding base for the organization is generally very restricted with the rare possibility of a temporary but very broad base from which funds might be had. They agreed that the diversity-of-funding-base parameter should be represented by a distribution built from the following coefficients.

Minimum = 1
Most likely = 2
Maximum = 8

A peakedness value of 5 was agreed upon because such a value creates a distribution shape that correctly represents their mental image regarding the relative frequencies of values near the most likely value and in the tails of the distribution. The resulting distribution is shown in Figure 2.17.

The level of funding for the organization has never been outstanding, but funds in the past have been made available for activities upon which the organization and host government agreed. In recent years, however, sympathetic government agents and bureaucracies have either been supplanted by less empathetic entities or have been coerced into more moderate attitudes with regard to funneling government

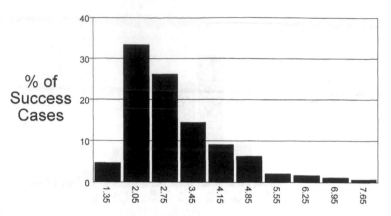

FIGURE 2.17 Distribution for diversity of funding base for terrorist group #2.

funds toward the terrorist group. As a result, the organization has had to adopt a much more moderate philosophy and attitude in target selection. The expense of extra-country operations has virtually precluded strikes outside the host-country's borders. In-country operations have been severely curtailed but not extinguished. It is the attitude of risk-team members that the government plan is to trickle just enough funds to the organization so that the government might have an "unauthorized" group to carry out its occasionally required dirty work. A short-lived infusion of significant funds could occasionally occur when the government deemed it necessary.

Risk-team members concluded that although the level of funding has decreased in recent years, it is in the government's best interest not to halt the flow of funds altogether. The agreed-upon funding-level distribution was built from the following coefficients.

Minimum = 2
Most likely = 4
Maximum = 7

A peakedness of 8 was assigned so that low but sufficient levels of funding would be emphasized. This distribution is shown in Figure 2.18.

The experience and technological prowess issue was next considered in the risk model. Level of experience of organization members was first addressed. As previously indicated, organization cells are scattered across the countryside. Each cell recruits its members from the local indigenous population. For any given member, cell membership is more a matter of prestige than one of political skullduggery or paroxysms of violence. Cells, therefore, tend to be populated primarily with inexperienced members or those who have played merely tangential roles in a multitude of campaigns. Some risk-team members pointed out, however, that members typically join a cell as adolescents and, by the time they could be considered for a critical role in a terrorist activity, they have a mindset and experience base that precludes labeling them as novices. In addition, each cell retains senior-level veterans, each with a long history

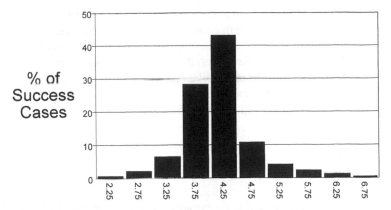

FIGURE 2.18 Distribution for level of funding for terrorist group #2.

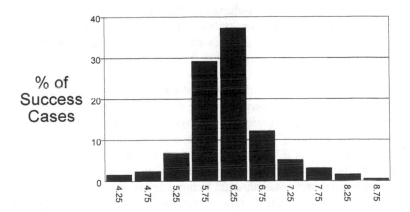

FIGURE 2.19 Distribution for level of member experience for terrorist group #2.

of successful escapades. Risk-team members, therefore, decided to characterize the experience level of members with a distribution built from the following coefficients.

Minimum = 4
Most likely = 6
Maximum = 9

A peakedness value of 8 was established so that the most likely value would be emphasized. This distribution appears in Figure 2.19.

Next discussed was effectiveness of execution. Although the majority of all members who actually participate in the organization's escapades is relatively inexperienced, the level of sophistication of their attacks is commensurate with their abilities. Organization-sanctioned activities tend toward the mundane — crude car-bomb attacks on easy-to-access and generally undefended targets, random slayings of unsuspecting tourists or commuters, and similar acts of low-tech mayhem. Due to the lack of sophistication and seemingly random nature of the attacks, cell members are uncannily successful in their execution, with only the occasional foible.

Risk-team members believed that the host government has had prior knowledge of many of the group's activities and could have curtailed a salient proportion of the attacks. Pressure on the government, it was reasoned, might significantly reduce the effectiveness of the organization. When all aspects had been considered, the risk team agreed that the effectiveness-of-execution parameter should be represented by a distribution built from the following coefficients.

Minimum = 5
Most likely = 7
Maximum = 9

A peakedness of 5 was agreed upon because values in the tails of the distribution needed to be stressed. A peakedness value of 5 creates a distribution shape that has frequencies near the most likely value and in the tails of the distribution that match the intuition of the group. The resulting distribution appears in Figure 2.20.

The final parameter to be considered in the relative-risk model was one that addresses the level of technical prowess exhibited by the organization. Given the organization being considered, risk-team members realized that this would be a complex issue.

Most attacks carried out by the organization are relatively low tech. There are several reasons for this. First, perpetrators of the attacks tend to be relatively young and unskilled cell members. In addition, the expense of high-tech weaponry generally precludes the acquisition of such arms by cash-strapped rural cells.

Local cells carry out terrorist attacks to further their own ends and to advance the agenda of the host government. Therefore, government-purchased and supplied high-tech munitions and commensurate training occasionally can find their way to strategically located rural cells. Corporate intelligence has learned, for example, that sophisticated shoulder-launched ground-to-air missiles (for use against aircraft) and

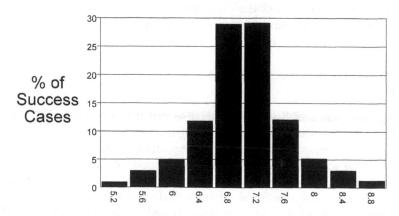

FIGURE 2.20 Distribution for effectiveness of execution for terrorist group #2.

ground-to-ground missiles (for use against armored vehicles) have, through middle-men, been funneled from the government to local cells. This occasionally is the situation when the government wishes the terrorist organization to carry out activities that the government dares not overtly execute.

Considering all of the facts, and after much debate, the risk team decided that the organization's level of technical prowess should be represented by a distribution built from the following parameters.

Minimum = 2
Most likely = 4
Maximum = 8

A peakedness value of 7 was established that will stress the value of 4, yet give significant emphasis to the upper range of the distribution. The resulting frequency plot is shown in Figure 2.21.

A final hurdle to risk-model execution was the establishment of weights for the three categories. Organizational aspects were first to be considered. Host-government support and political backing is essential to the existence of most cells and certainly is vital to the orchestration of the disparate cells by the organization leader. Risk-team members, therefore, harbor little doubt that cataclysmic degradation in organization might be affected by bringing to bear political and economic pressure on the host government.

Similarly, risk-team members theorized that funding for the organization might significantly be curtailed if the threat were made to withdraw business and to expose host-government/terrorist-organization ties. Team members decided that the corporation would assign weights of relatively great magnitude to the organizational and funding categories because expenditure of corporate funds in these areas has the potential for measurable impact.

Conversely, team members felt much less confident with regard to reducing the organization's technical prowess. First, they reasoned, the level of technical

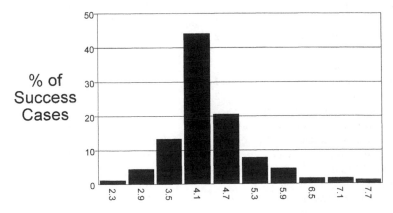

FIGURE 2.21 Distribution for level of technical prowess for terrorist group #2.

capability of most cells already is fairly meager. Significant reduction in this already paltry capability was deemed unlikely. In addition, team members reasoned that when the host-government officials felt strongly enough about a situation to prompt them to supply the organization with high-tech weapons, there would be little the corporation could do to counter these sporadic and well-concealed events. It was agreed, therefore, that the weight for this category would be of relatively small magnitude.

Weights of 0.4, 0.4, and 0.2 were entered in the model, respectively, for the organizational, funding, and experience/technical prowess categories. The results from model execution is delineated in the following section. A summary of model input parameters appears in Table 2.2.

RELATIVE-RISK MODEL RESULTS FROM EVALUATION OF TERRORIST ORGANIZATION #2

Results from application of the risk model to data relevant to this terrorist organization are shown in Figures 2.22 through 2.25. Figure 2.25 indicates that the modal (and mean) weighted total risk value for terrorist organization #2 is around 4.1. The minimum weighted value is about 3.1 and the maximum is about 5.6. It was observed that the distributions that represent organizational and funding parameters stress coefficients of relatively low magnitude, while coefficients for technological prowess are of relatively great magnitude. However, the coefficients that comprise distributions representing entire organizational and funding categories are of relatively great magnitude when compared with those that constitute the technological prowess

TABLE 2.2
Input Parameters for Relative-Risk Model — Terrorist Organization #2

Category/Parameter	Minimum	Most Likely	Maximum	Peakedness
Organization Category				
Level of Organization	1	2.5	9	3
Size of Organization	1	3	9	8
Logistical Support from Host Government	2	3.5	8	4
Funding Category				
Diversity of Funding Base	1	2	8	5
Level of Funding	2	4	7	8
Experience/Technology Category				
Level of Experience of Members	4	6	9	8
Effectiveness of Execution	5	7	9	5
Level of Technological Prowess	2	4	8	7
Category Weights				
Organization Category Weight		0.4		
Funding Category Weight		0.4		
Experience/Technology Weight		0.2		

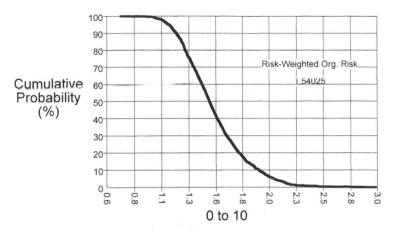

FIGURE 2.22 Cumulative frequency curve showing range of weighted risk associated with organizational aspects of terrorist group #2.

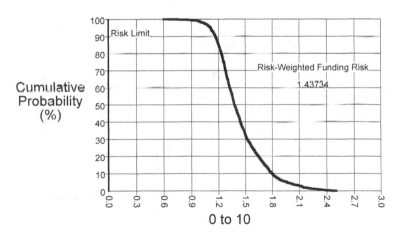

FIGURE 2.23 Cumulative frequency curve showing range of weighted risk associated with funding aspects of terrorist group #2.

category distribution (Figures 2.22, 2.23, and 2.24). Again, of course, this is a function of the relatively great weight given to the first two categories.

COMPARISON OF THE TWO TERRORIST ORGANIZATIONS

Risk-team members realized that in the first fiscal year the corporation would be financially capable of attempting to establish a presence in only one country. They realized, therefore, that they would need to compare the results from the two completed analyses so that they could go forward with a single recommendation. It was decided that they would use as input the Weighted Total Risk curves from

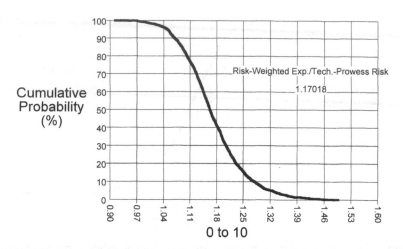

FIGURE 2.24 Cumulative frequency curve showing range of weighted risk associated with experience/technological prowess aspects of terrorist group #2.

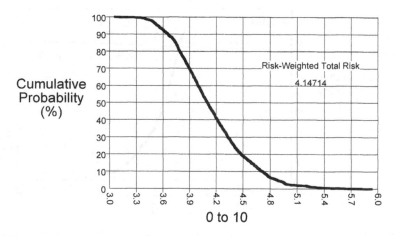

FIGURE 2.25 Cumulative frequency curve showing range of total weighted risk associated with terrorist group #2.

both analyses. In the new model, one curve would be subtracted from the other to result in a "difference" curve. A "difference" curve could also be generated for each category.

To the relative-risk model, the following lines of code were added to facilitate the passing of the output-risk curves to files on the computer's hard disk.

```
SAVE wt_org_risk c:\nb\to1org.txt;
SAVE wt_fund_risk c:\nb\to1fund.txt;
SAVE wt_tech_risk c:\nb\to1tech.txt;
SAVE wt_total_risk c:\nb\to1total.txt;
```

where

SAVE is a system command that saves distribution values to a file, in the order they are generated by the Monte Carlo process

wt_org_risk is the organizational weighted risk output variable

c:\nb is the hard drive and directory name where the output distributions will be stored

to1org.txt is the name given to the file that will store the 1500 values that make up the organizational output distribution for terrorist organization #1

wt_fund_risk is the funding weighted risk output variable

to1fund.txt is the name given to the file that will store the 1500 values that make up the funding output distribution for terrorist organization #1

wt_tech_risk is the technical prowess weighted risk output variable

to1tech.txt is the name given to the file that will store the 1500 values that make up the technical prowess output distribution for terrorist organization #1

wt_total_risk is the total weighted risk output variable

to1total.txt is the name given to the file that will store the 1500 values that make up the total-risk output distribution for terrorist organization #1

The new risk program that calculates the difference between output distributions from the two analyses contained the following lines of code.

```
orgdiff = to1org – to2org;
funddiff = to1fund – to2fund;
techdiff = to1tech – to2tech;
totaldiff = to1total – to2total;
```

where

orgdiff is a variable that holds organization-related differences between terrorist organizations 1 and 2

to1org and **to2org** are the variable names that hold the data contained in the to1org.txt and to2org.txt files, respectively

funddiff is a variable that holds the funding-related differences between terrorist organizations 1 and 2

to1fund and **to2fund** are the variable names that hold the data contained in the to1fund.txt and to2fund.txt files, respectively

techdiff is a variable that holds the technology-related differences between terrorist organizations 1 and 2

to1tech and **to2tech** are the variable names that hold the data contained in the to1tech.txt and to2tech.txt files, respectively

totaldiff is a variable that holds the total risk-related differences between terrorist organizations 1 and 2

to1total and **to2total** are the variable names that hold the data contained in the to1total.txt and to2total.txt files, respectively

The plots for the output variables orgdiff, funddiff, techdiff, and totaldiff are shown in Figures 2.26 through 2.29. It can be seen from the plots in Figures 2.26

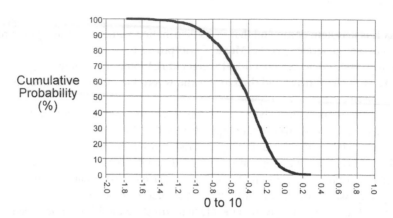

FIGURE 2.26 Cumulative frequency curve showing range of difference between organizationally related weighted risk for terrorist groups 1 and 2.

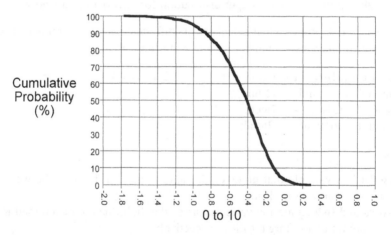

FIGURE 2.27 Cumulative frequency curve showing range of difference between funding-related weighted risk for terrorist groups 1 and 2.

and 2.27 that the organizational and funding aspects for terrorist organization #2 are generally more risky than those for organization #1, but only slightly so as indicated by the relatively small magnitude of the negative values. In the Experience/Technical Prowess category difference plot (Figure 2.28), all X-axis values are of relatively great magnitude and are positive. This signifies that for this category, terrorist organization #1 is significantly more risky than organization #2. This relatively great risk is reflected in the Total Risk difference plot (Figure 2.29), in which most of the values are positive.

The plot (Figure 2.29) indicates that 90% of the time, terrorist organization #1 represents a risk that is 1.5 times greater than that posed by organization #2. The team, therefore, concluded that they would recommend that the corporation first attempt to do business in the country in which organization #2 operates. The next

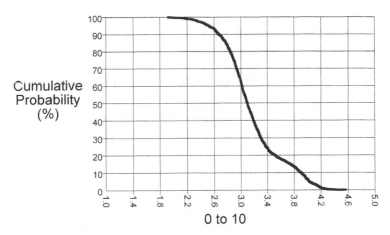

FIGURE 2.28 Cumulative frequency curve showing range of difference between experience/technical prowess-related weighted risk for terrorist groups 1 and 2.

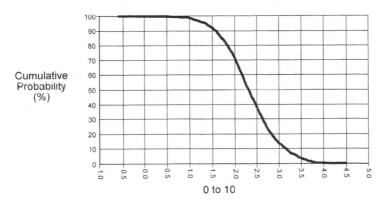

FIGURE 2.29 Cumulative frequency curve showing range of difference between total weighted risk for terrorist groups 1 and 2.

task was to determine just how expensive such a venture might be. To determine the cost, an absolute-cost risk model was employed.

BUILDING THE TERRORISM ABSOLUTE-COST RISK MODEL

It can be a daunting task to communicate risk modeling results to decision makers who are not experts in risk/uncertainty. This consideration served as the guiding light for the risk-team's decisions regarding development of the terrorism absolute-cost risk model.

Although a truly comprehensive cost model might include many tangible and intangible costs, the team decided that ease of presentation, simplicity, and consistency were of utmost importance. They therefore decided to build an absolute-cost model that would emulate the relative-risk model by addressing the same categories.

The new model, however, would require as input absolute dollar amounts to build the distributions.

Risk-team members realized that expenses to mitigate terrorist activities would be ongoing; that is, the corporation would incur costs in all years of operations. Some costs, such as building facilities, would be most severe in initial years. Other costs, however, such as providing increasing levels of diplomatic services, would increase with time. Although risk-model builders understand that costs per year per parameter can be estimated and entered into a time-series risk model so that a Net Present Value (cost) can be calculated, it was felt that for a 10-year model, to get 10 distributions per parameter from the suppliers of cost information was unrealistic. In addition, it was reasoned that a presentation of fewer numbers would increase their chance of making their point — fewer points to explain and about which to argue. Therefore, the risk team opted to forgo the time-series approach for a simpler model in which the values entered represent best guesses at average yearly costs.

In the organization category, the following parameters apply.

- Cost to reduce level of organization
- Cost to reduce size of the terrorist group
- Cost to reduce logistical support from the host government

Risk-team members agreed that the definition of *Cost to reduce level of organization* would be the cost to cause a significant, measurable, and meaningful reduction in the communication and coordination among disparate entities within the organization. Similarly, they agreed that *Cost to reduce size of the terrorist group* would be defined to be the cost to reduce the size of the organization to a level that would measurably decrease the possibility of a negative impact on corporate operations. The parameter *Cost to reduce logistical support from the host government* was agreed to mean the cost to influence the government in such a way as to cause a significant and meaningful reduction of support for the terrorist organization.

The following funding parameters were agreed upon.

- Cost to reduce the number of funding sources
- Cost to reduce overall level of funding

Cost to reduce the number of funding sources was defined as the amount the company should spend to convince current funding entities to cease support for the organization and to persuade potential sources to refrain from funding the organization. *Cost to reduce the overall level of funding* was defined as the expense to the corporation to make sure that entities, such as the host government, do not attempt to compensate for lost revenues from other various sources or to increase their overall budget for the organization.

The technical-prowess category contains the following three parameters.

- Cost to reduce organization-member experience level
- Cost to reduce the effectiveness of the organization
- Cost to reduce the organization's technical capabilities

Cost to reduce organization-member experience level was defined as the cost to the corporation to entice more experienced members to leave the organization and to reduce training and other experience-yielding activities for remaining members. The *Cost to reduce the effectiveness of the organization* is the expense to the corporation to learn about and disrupt terrorist activities. The *Cost to reduce the organization's technical capabilities* is the corporate expense associated with curtailing the flow of high-tech weapons to the organization.

This model will calculate absolute dollar amounts. Reluctance to deem as important a particular variable or category can be reflected in the amount one is willing to spend on it. Therefore, it was decided to forgo the application of category weights in this model.

ABSOLUTE-COST RISK MODEL EQUATIONS

Because the absolute-cost risk model will result in simple probabilistic sums of all costs, the equations are simple and straightforward. They are

OrgTotal = OrgLevel + SizeOrg + Support;
FundTotal = FundSources + OverallFund;
TechTotal = Experience + Effect + Tech;
TotalCost = OrgTotal + FundTotal + TechTotal;

where

OrgTotal is a variable that holds the sum of all organizationally related costs
OrgLevel is the cost to reduce the organization level of the terrorist group
SizeOrg is the cost to reduce the organization's membership total
Support is the cost to dissuade the host government from supporting the organization
FundTotal is a variable that holds the sum of all funding-related costs
FundSources is the cost to reduce the number of funding sources for the organization
OverallFund is the cost to reduce the overall organization budget
TechTotal is a variable that holds the sum of all experience/technical-prowess-related costs
Experience is the cost to curtail the level of individual-member experience
Effect is the cost to reduce the organization's effectiveness
Tech is the cost to diminish the organization's ability to obtain, deploy, and utilize high-tech weaponry and tactics
TotalCost is a variable that holds the total mitigation cost

APPLICATION OF THE ABSOLUTE-COST RISK MODEL TO TERRORIST ORGANIZATION #2

First to be debated were the costs associated with reducing the terrorist group's ability to organize and communicate. Corporate terrorism experts agree that most terrorism-cell members, like gang members in the United States, join local cells due to lack of meaningful and gainful employment in the area. Because the geographic area of the host country is not great, corporate planners are considering a scheme in which they will establish many small manufacturing, packaging, and distribution facilities across the country. Known cell members would be actively recruited to work at and draw salaries from local facilities. The hope is to foster a dependence on the income gained and thereby acquire an interest by local cell members in keeping the corporation operating.

In addition, the corporation plans to respond to the first terrorist attack of any kind anywhere by shutting down all operations at all regional facilities. These shutdowns will be of sufficient length to seriously impact incomes of local employees. The hope is that cell members from around the country will bring pressure to bear on the offending party and convince it to cease attacks on corporate properties.

Corporate planners believe that the yearly average cost of a centralized facility, added to the cost of at least one 2-week shutdown of all facilities would be represented by the following distribution-building parameters.

Minimum = $2.7M
Most likely = $5M
Maximum = $7.3M

A peakedness value of 5 was established so that the most likely value would be sufficiently stressed while giving values in the tails of the distribution sufficient representation. The resulting distribution is shown in Figure 2.30.

Risk-team members agreed that reducing the size of the organization would be a tall order. Although they hoped to curtail attacks by offering corporate employment to cell members, risk-team members harbored no unrealistic expectations with regard to a correlation between increased cell-member employment and a reduction of cell

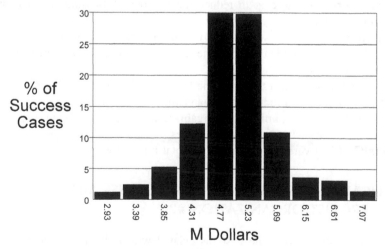

FIGURE 2.30 Distribution for cost to reduce organization-level risk for terrorist group #2.

membership. However, one possible ploy would be to publish nationwide the names of known and suspected contributors after execution of a terrorist activity. The corporation could also express that it is considering making a condition of employment that employees not be associated in any way with terrorist organizations. It is envisioned that public exposure, disruption of income, and the threat of being fired because of terrorist ties will, over time, significantly reduce organization membership. The following costs of implementing an advertising campaign for a year are anticipated.

Minimum = $0.5M
Most likely = $1.0M
Maximum = $2.0M

A peakedness of 10 was assigned to reflect the fact that such a campaign is expected to cost around $1M. The resulting distribution is shown in Figure 2.31.

Because the corporation will operate in the country and will contribute significantly to the tax base, it is hoped that corporate executives will realize some influence with the host government with regard to reducing support for terrorist activities. Corporate executives are fairly confident that they can reduce direct attacks on corporate facilities. However, the government is likely to continue to fund terrorist attacks both on anti-government groups within the country and against less-than-friendly neighboring states. Supply routes for corporate facilities could be disrupted by attacks on neighboring countries and by damage to local supply lines.

The corporation, therefore, is considering the establishment of a permanently staffed conference center where conflicting factions would be invited, at corporate expense, to meet and attempt to settle their differences by nonviolent means. A permanent staff or corporate-employed "ambassadors" would provide "shuttle diplomacy" when appropriate. The yearly cost of such a facility and staff is expected to be (after facility construction)

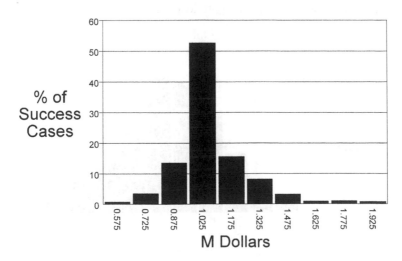

FIGURE 2.31 Distribution for cost to reduce organization-size risk for terrorist group #2.

Minimum = $3M
Most likely = $3.7M
Maximum = $4.5M

A peakedness of 2 was established to indicate the relative indecision by team members with regard to costs. A plot of the resulting distribution is shown in Figure 2.32.

Reduction of the number of funding sources was next considered. The organization is primarily funded by the host government, but some influx of funds does come from sympathetic bordering and geographically removed countries. Corporate planners reason that they might offer fund-contributing countries a "piece of the action" in return for a halt in terrorist support. Such a "piece of the action" might consist of directing a supply route through such a country rather than where the supply line might most economically be established. Corporate planners expect that the average yearly cost of such concessions might be

Minimum = $5M
Most likely = $10M
Maximum = $25M

A peakedness of 2 was considered appropriate because of the uncertainty of planners regarding these costs. The distribution is shown in Figure 2.33.

Corporate planners realized that shortfalls in terrorist organization coffers due to the withdrawal of contributions from foreign entities will likely be countered by larger infusions of cash from the host government. This would primarily take place when the government wished to use the terrorist organization for a covert government-sanctioned terrorist activity. Such activities are initiated by a small cadre of government officials who still view acts of terrorism as justifiable. The corporation can do nothing to overtly attempt to remove these individuals from office. However, the corporation can contribute, ndirectly, to the election campaigns of ithose attempting to depose such undesirable individuals. Corporate executives estimate that the average yearly cost of such funding likely would be

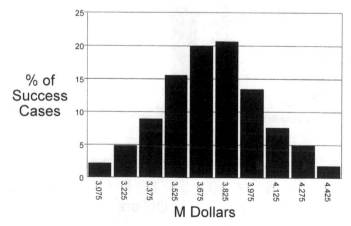

FIGURE 2.32 Distribution for cost to reduce logistical-support risk for terrorist group #2.

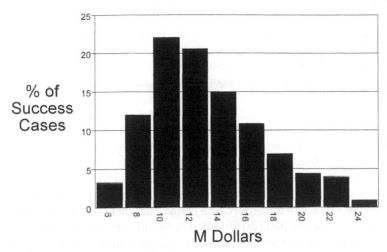

FIGURE 2.33 Distribution for cost to reduce funding-source risk for terrorist group #2.

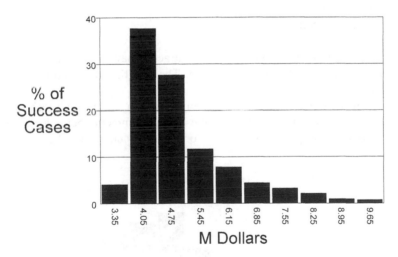

FIGURE 2.34 Distribution for cost to reduce overall funding risk for terrorist group #2.

Minimum = $3M
Most likely = $4M
Maximum = $10M

A peakedness of 7 was assigned to emphasize the lower end of the range. The resulting distribution is shown in Figure 2.34.

A reduction of the experience level of individual cell members, it is reasoned, will follow as more time is spent by cell members in gainful employment activities. However, to have an impact on experience, corporate executives realize that camps that serve as terrorist-training centers must be closed. These camps are located in remote and relatively inaccessible areas of the country; therefore, any plan to convert these facilities to more benign corporate use would be unrealistic. Corporate officials reason that the only way to close the bases is to pay the government to do so and then to monitor activities at the remote locations. The yearly average cost of such a buy-out and monitoring plan is expected to be

Minimum = $10M
Most likely = $15M
Maximum = $20M

A peakedess of 1 was assigned to indicate the low confidence in the most likely value. The distribution of costs is shown in Figure 2.35.

Corporate officers surmise that there is nothing the corporation can do to reduce the effectiveness of terrorist attacks on noncorporate facilities. However, every critical corporate office will effectively be a compound, including secured housing for permanent and visiting staff, expatriate security officers, armed guards, extensive camera systems, physical barriers, and other practical and sophisticated measures. Executives estimate the average yearly cost of such measures to be

Minimum = $5M
Most likely = $7M
Maximum = $7.5M

Because bids from security agencies and contractors have already been procured, a peakedness of 10 was deemed appropriate. The resulting distribution is shown in Figure 2.36.

The terrorist organization is a relatively low-tech outfit. High-tech munitions and weaponry are obtained only when the host government deems it necessary to supply such arms for specific purposes. Corporate executives feel there is little they

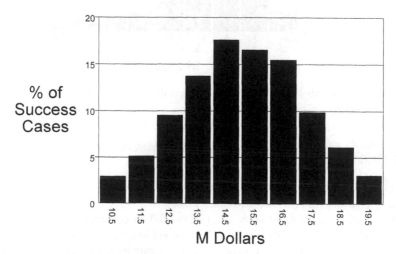

FIGURE 2.35 Distribution for cost to reduce member-experience risk for terrorist group #2.

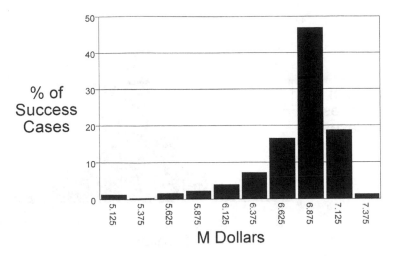

FIGURE 2.36 Distribution for cost to reduce organization-effectiveness risk for terrorist group #2.

can do, above and beyond what has already been outlined, to stop sporadic influxes of high-tech arms to the organization. Therefore, the budget for such activities was given the token amounts

Minimum = $0.5M
Most likely = $0.8M
Maximum = $1.0M

A peakedness of 5 was established. The resulting distribution is shown in Figure 2.37. Table 2.3 shows a summary of all input values to the absolute risk model.

ABSOLUTE-COST RISK MODEL RESULTS FOR EVALUATION OF TERRORIST ORGANIZATION #2

Results from execution of the absolute-cost risk model are shown in Figures 2.38 through 2.41. Costs to battle organizational aspects are least. The organization-category costs range from about $6.7M to $13M with a modal value of about $10M. The price to counter funding and technical-prowess attributes is of significantly greater magnitude. The price to mitigate funding considerations ranges from $8M to $30M with a modal value around $19M. Implementing technology-fighting policies will cost the corporation between $16M and $27M.

The total average cost per year for the corporation will range from $38M to $66M with a most likely price of about $49M. As can be seen from Figure 2.41, there is a 90% chance that the terrorist-fighting price will be about $44M or more, a 50% chance that costs will exceed about $50M, and only about a 10% chance that more than about $56M per year will need to be spent.

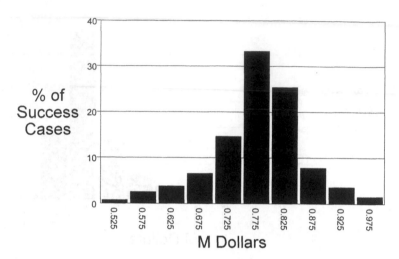

FIGURE 2.37   Distribution for cost to reduce technical-prowess risk for terrorist group #2.

TABLE 2.3
Input Parameters for Absolute Risk Model — Terrorist Organization #2

Category/Parameter	Minimum	Most Likely	Maximum	Peakedness
Organization Category				
Cost to Reduce Organization Level ($Millions)	2.7	5	7.3	5
Cost to Reduce Organization Size ($Millions)	0.5	1	2	10
Cost to Reduce Logistical Support ($Millions)	3	3.7	4.5	2
Funding Category				
Cost to Reduce Funding Sources ($Millions)	5	10	25	2
Cost to Reduce Overall Funding ($Millions)	3	4	10	7
Experience/Technology Category				
Cost to Reduce Member Experience ($Millions)	10	15	20	1
Cost to Reduce Org. Effectiveness ($Millions)	5	7	7.5	10
Cost to Reduce Tech. Capabilities ($Millions)	0.5	0.8	1	5

SO, SO WHAT?

Well, we certainly have gone through a lot of trouble to this point. So, why are we better off for having done this?

A major benefit of any risk-model undertaking is that it forces participants to agree on just what question is to be answered. This is often one of the more difficult goals to achieve. Much of the heated discussion that ensues in a poorly run and facilitated risk-modeling meeting (or any type of meeting, for that matter) is a result of different people arguing for the resolution of problems that are not only different, but often at odds with one another. Gaining clarity and consensus, up front, of just what problem needs resolution makes all the difference. The fact that the group

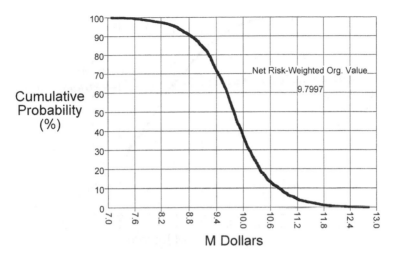

FIGURE 2.38 Cumulative frequency curve for cost to reduce overall organizational risk for terrorist group #2.

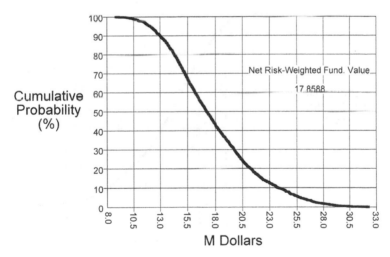

FIGURE 2.39 Cumulative frequency curve for cost to reduce overall funding risk for terrorist group #2.

could decide upon the type of model (Monte Carlo vs. a decision-tree or a linear-programming model) was paramount. In addition, agreement on the need for relative- and absolute-model types and agreement on the parameters and parameter definitions for each model are significant and necessary steps. Getting all participants "on the same page" is a difficult but fundamental process.

It should be noted that there is nothing special about the risk models presented here. The variables used and the code displayed are but one set of variables and

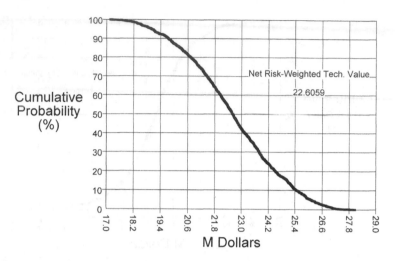

FIGURE 2.40 Cumulative frequency curve for cost to reduce overall experience/technical-prowess risk for terrorist group #2.

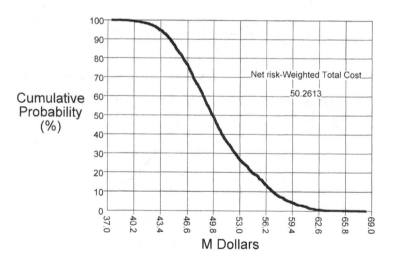

FIGURE 2.41 Cumulative frequency curve for cost to reduce overall risk for terrorist group #2.

programs that might have been used to attack this problem. These models are not presented as "the way to do it"; rather, these models are meant to instill in the reader that "it can be done." This is just one way, and not necessarily the best way, to accomplish the task.

Validation of the models can be done only over time. Attempting to validate the model relative to some other path that might have been taken or model that might have been used is folly. You rarely, if ever, have the opportunity to know the result(s) of the "road not taken." In the case of the company described here, the only validation possible is to deem whether the decisions made, using model output, were, in the end, good decisions for the company. If the majority of the decisions made result in economic prosperity for the company, then the models are vindicated on those grounds. You may argue that the company might have done just as well or better had it not even considered undertaking the risk-model route. You are welcome to do that.

An associated benefit is the conversations that are necessary to populate the model with data. Whether discussing relative rankings or actual costs, members of the team must engage in in-depth and meaningful conversations concerning each parameter for each country. Many times, the benefit of these conversations outweighs the benefit of actually executing the model and interpreting the results.

The relative-risk model is one of consensus and is generic in nature. This means that the group can use the model time and again to evaluate new countries or to reevaluate countries over time. The fact that the answer variable for the model is predicated on a predetermined scale (in this case, 0 to 10) facilitates comparison and ranking of all countries evaluated by the model.

Output from the absolute model gives management a means to view the range of expenses associated with a given mitigation strategy and the associated probabilities. Models like this rarely make clear what decision should be taken; however, the data generated by such models are essential inputs to the decision-making process. For example, the subject group of this chapter might have selected terrorist organization #2 because, by the output of the relative-risk model, it seemed to be the better bet. However, using the absolute-risk model, management might view the costs associated with mitigating problems associated with terrorist organization #2 and decide that the costs are prohibitive. Running the absolute-risk model on terrorist organization #1 might reveal that although it had a higher relative "risk," the cost of mitigation of those risks is lower. Combination of the outputs from both models, in this way, allows companies to make better decisions.

3 Gathering Information Consistently in an Inconsistent World

CONTENTS

INTRODUCTION

In several respects, this chapter differs from the others in this volume. The preponderance of chapters that compose the Examples section of this book demonstrate how a stochastic model or combination of models can be employed as the solution to business and other problems. As elucidated in the first book (*Risk Assessment and Decision Making in Business and Industry: A Practical Guide*), the steps of building and executing the risk model are ancillary to the organizational, political, human, and other problems with which one must first grapple.

In the first book, an entire chapter entitled "Consistency — The Key to Risk Assessment" was devoted to the concept of consistency. As asserted in that chapter, except in "one-off" situations, maximum benefit from risk processes is realized only when multiple opportunities are consistently assessed or compared. In the real world, the problem of achieving consistency can be a tough nut to crack.

Consistency comes in a variety of flavors. The following are some of the salient aspects of consistency.

- Building "type" consensus models
- Embracing a common measure
- Stressing consistency over being "right"
- Preventing practitioners from "gaming" the system
- Locking models
- Databasing inputs and outputs
- Risk-assessment vision

The practice of building "type" consensus models centers on the concept of assessing the maximum number of opportunities/liabilities with the smallest (practical) number of risk models. Not only should the number be small, but each model should have been arrived at by consensus of those who will put it to use. In addition, embracing a common measure enforces the premise that unless various entities are considered on a common scale (dollars, percent return, etc.), attempts to compare, rank, or portfolio-manage the items are folly.

For situations in which comparison and/or ranking is paramount, being consistent might be much more important than being "right." For example, if the task is to rank-order a number of objects by weight, it does not matter whether each object's weight was measured correctly, but only that the measurements were consistent so ranking is possible. In business, this can be a powerful concept.

Regardless of the sophistication built into a risk model or process, it is only a matter of time before users of the process or model figure out how to "game" the system; that is, users deduce how to manipulate the system so that system output favors their proposed project. In most companies, government institutions, and academic arenas, there is intense pressure to cut costs or, in some other way, make the proposal appear to be the most economically reasonable choice among many competing options. Therefore, there exists an overwhelming compulsion to generate a risk assessment result that casts a favorable light on the proposal.

In my years of traversing the globe and observing methods to prevent "gaming" of the system, I have noted two dominant approaches to dealing with this problem: the "police" method and the "accountability" process.

Under the "police" method, organizations attempt to curtail manipulation of the system by establishing a "risk police squad" (RPS). RPS membership typically includes senior members of the staff who have much experience in a particular area. Just one RPS responsibility is to review all potential risk-model inputs prior to model execution. In this scheme, anyone who wants a project and associated assessment results considered for funding, for example, must have the model input data formally reviewed by the RPS. In a review, it is the responsibility of RPS members to attest to the fact that input parameters are reasonable.

The "police" approach embodies several advantages and drawbacks. A distinct advantage is that this approach does, in fact, keep overly optimistic or overly pessimistic or otherwise bizarre data from being considered by the model. Another real benefit is that it forces discussion of the project by a group that is not intimately involved with the project. Such discussions often are very valuable for the project proponents.

A distinct disadvantage is that this method of enforcing consistency is often perceived (and rightly so, in most cases) as an indication of lack of trust. It also is taken as a "big brother" tactic that imbues the organization with an "us vs. them" attitude. Another major drawback of this process is that it is time consuming. Projects can only move forward at the rate at which they can be reviewed. Often, scheduling a review session with the RPS can be almost as daunting a task as reviewing the project.

The "accountability" method exhibits its own set of advantages and drawbacks. Under this scheme, project proponents are absolutely held accountable for achieving the results that they propose their project will generate. Severe consequences will be swift and sure if projections used to sell the project are not met.

One advantage to the "accountability" system is that it certainly squelches any desire to create or espouse unachievable expectations on the part of project proponents. This system also allows projects to proceed at the most expedient rate. Projects do not consume corporate resources other than those directly related to the project.

One glaring disadvantage to this approach is that it creates a workforce that is "running scared." When severe consequences result from missed objectives, managers of such organizations will find that project proponents will "set their sights low" so that they know objectives can be achieved. Major advances typically are only made when "stretch goals" are set; that is, when the bar is set high. High bars are rare in such organizations. In addition, some of the most valuable employees are those with a wide range of experiences — including failure. Existence of such individuals might not be commonplace in entities that strictly employ the "accountability" process.

Yet another means of achieving consistency in assessment is that of locking the models; that is, generating a model that cannot be modified by the end user except by entering data. This tactic is often employed when multiple opportunities in farflung geographic locations have to be consistently assessed. Anyone who has, for example, distributed a spreadsheet to various locations with the idea that users will simply enter data and execute the sheet does not really understand users. Before you know it, there are as many different versions of the spreadsheet as there are users to whom it was sent. Often, the modifications made to the model preclude comparison of outputs from the model. The practice of distributing "enter-data-and-run-only" models helps avoid the multiple-version syndrome.

Although sending to a database the input and output parameters from a risk model nearly always is sound practice, the facet of consistency related to storing risk-related information in a database is often overlooked. The act of stashing away data in an electronic repository does not necessarily enhance or promote consistent behavior. This is especially true if just anyone can store any data preferred. For maximum benefit, the data stored should be at least those from sanitized or documented analyses, or at best, data from analyses that have been post-appraised and confirmed/validated. When such data are available, they can be a powerful guide for input data related to new opportunities. In real time and in reaction to a user's input values, some companies have gone so far as to query the database and display how well a user's input value for a parameter "fits" with historical data from similar projects.

All aspects of consistency related to risk and uncertainty analyses are much more easily communicated to, and accepted by, an organization when those aspects are clearly seen as integral and necessary components of a vividly described and communicated risk vision. For example, a CEO or subordinate might describe with text and figures that, in the near future, the company would like to be able to portfolio-manage its wide variety of opportunities on the basis of risk-weighted value. The vision might be for each opportunity within the corporation to have a risk/uncertainty assessment performed on it that culminates in an expression of risk-weighted value to the corporation. The communication might further point out the real benefits of "doing the right project" and the financial gains from such an approach. Acceptance and understanding of the ultimate goal (vision) by corporate personnel always makes easier the implementation of the many cultural, political, and organizational changes that will be required to realize the vision. Without understanding and acceptance of the vision, mandated changes in routines generally are viewed as unnecessary and wasteful stirring of the pot.

In this chapter the following very real and vexing problems will be addressed.

- Needing to use a single risk model to assess multiple but similar entities
- Recognizing the differences between entities that are deemed to be similar in type
- Devising a methodology that accounts for those differences, yet allows collection of model-input data in a consistent manner

Although the solution to the problem posed in this chapter is a solution generally employed within the probabilistic world, I first witnessed its application at my former company in relation to assessing and ranking production opportunities. My good friend and former colleague Kim Overcash elegantly employed this solution as part of his work in AEGNA (Amoco Energy Group North America), and recognition is here given to Kim for his pioneering efforts.

THE PROBLEM

Don is a site engineer at a major manufacturing facility. For the last 20 years he has been in charge of waste cleanup at the plant. Don has done such a stellar job that corporate officials have relegated to him the unenviable task of coordinating such cleanup efforts at several other corporate facilities.

Even though Don's job is concerned with environmental issues, Don has always been aware that "the bottom line" is still important; that is, the cost of a project will dictate the extent to which a site is cleaned up. Therefore, Don's probabilistic model has always included a heavily weighted cost component. Don's initial task is to select sites to be cleaned up and to determine the associated site-evaluation costs exclusive of actual cleanup costs. As he had done at his home facility, Don will then submit his findings to management which will integrate his data with other information to determine which sites will first be addressed and to what extent. Only then can actual cleanup costs be determined. For his home site, Don gathered data on the items in Table 3.1 to determine the cost of deciding which sites would even be considered for cleanup and what the costs of initial evaluation of those sites would be.

Although he was showered with kudos for his cleanup efforts at his home facility, Don deemed his part of the cleanup project a "cake walk." Because of his 20 years at the plant, he knows "everybody and everything" about that facility. Therefore, Don was quite shocked when he attempted to implement his data-gathering process at his first away-from-home facility.

Don traveled first to a west-coast facility where Phil played a role similar to Don's at the east-coast plant. At an initial meeting, Phil and Don reviewed Don's list of items for data gathering. Don immediately became aware that Phil wanted to discuss concerns that Don had either never considered or never consciously taken into account.

TABLE 3.1
Model Parameters

Site selection
Vendor selection
Number of plant personnel on project over 1 year
Number of vendor personnel on project over 1 year
Cost per person for plant personnel
Cost per person for vendor personnel
Selection of sampling techniques
Number of samples taken over 1 year
Cost per sample taken
Cost per sample analysis
Cost for sample-analysis interpretation

For example, for the very first item — Site selection — Phil insisted on discussing how weather would be a major factor. Because Don's facility is in the Northeast and subject to snowstorms and other inclement meteorological phenomena, all of the cleanup sites at Don's facility were either inside tanks, underground, or in some other way sheltered from the elements. At Phil's facility on the mainly sunny west coast, many of the cleanup sites are exposed to the weather. Inclement conditions are not often encountered, but favorable conditions certainly cannot be counted upon to prevail. Therefore, the effects of weather had to be considered.

In addition, Phil wanted to discuss the financial and political consequences of using unionized labor. At Don's facility, labor unions had fallen into disfavor a decade or so ago when the Northeast experienced a severe decline in heavy industry and manufacturing. None of the personnel at Don's facility belong to a union and very few vendors are unionized. At Phil's facility, however, about half of the plant personnel belong to unions. This also is true of vendors.

Because of Don's familiarity with the situation at his home facility, he realized that, although he had subconsciously taken into account many factors when he considered the cost of an individual item on his list, other people, like Phil, had very different concerns and insisted on having those concerns recognized and accounted for in the cost assessment. After visiting with Phil, Don paid similar visits to two more facilities and then returned home with his findings.

It was clear to him that site-specific concerns had to be considered in the model. It was also clear that it would be impractical to build a model that incorporated dozens of variables, some of which would apply to only a single site. Because the computer-based risk model was to be distributed to each site for execution by local personnel, it would be detrimental to acceptance of the model to include up to 100 parameters, only a small fraction of which would be recognized and employed at any given site. It also would not be practical to consider creating multiple site-specific models. The aim of the entire exercise was to provide a consistent set of data so management could rank and prioritize the corporation's cleanup efforts.

Results from multiple site-specific models might not facilitate comparison of results, and, besides, Don did not have the time to generate multiple models.

THE SOLUTION

Don could not help but believe that he was not the first person to encounter this dilemma. He decided to take the bold step of traveling to corporate headquarters where experts in such things reside. An initial email exercise connected Don with Kim, the corporate guru regarding probabilistic assessments. Don traveled to the headquarters location for a meeting with Kim. At the meeting, Don delineated to Kim his seemingly insurmountable problems. Kim had encountered exactly this difficulty many times and suggested a workable solution.

Kim's solution involved the creation of a matrix. He suggested that in the left-hand vertical column, he and Don list generic items that were common to all potential cleanup sites. This first column would be titled Generic Factors. The next 5 columns should be reserved for collecting probabilistic information for each generic item, such as minimum cost, most likely cost, maximum cost, peakedness, and percent probability. Other column headings would reflect the concerns for individual sites or letters or numbers that pointed to items in a table that listed such items.

Before Don returned home to implement this plan, Kim and Don created an example of such a matrix using a subset of Don's variables. Because Don had previously successfully implemented a cleanup assessment and because he subconsciously accounted for many site-specific factors, the duo agreed that Don's initial list of model parameters would serve as a good first attempt at creating a list of generic factors. Table 3.2 is a subset of site-specific factors that Don had collected during his visits.

TABLE 3.2
Site-Specific Concerns

A. Weather
B. Labor unions
C. Sampling permits
D. Site-specific corporate policies
E. Long-distance shipment of samples
F. Shallow bedrock
G. Nearby streams or rivers
H. Significant slope to topography
I. Nearby neighbors

The sample matrix is shown in Figure 3.1. Using this matrix as an example, Kim suggested that Don send out such a matrix to each facility along with a table that related the A through I column headings to a list of site-specific concerns. Then, for each generic item, personnel at a given facility could decide which of the site-specific items impacted its estimate. Each item considered to impact a generic parameter

GENERIC FACTORS	MINIMUM	MOST LIKELY	MAXIMUM	PEAKEDNESS	PROBABILITY	A	B	C	D	E	F	G	H	I
Site selection ($)														
Vendor Selection ($)														
# Plant Personnel/Yr. (#)														
Cost/Plant Person 1 Yr. ($)														
Cost/Vendor Person 1 Yr. ($)														
# Vendor Personnel/Yr. (#)														
Selection of Sampling Tech. ($)														
# Samples Taken/Yr. (#)														
Cost/Sample ($)														
Analysis Cost/Sample ($)														
Cost of Analysis Interp. ($)														

FIGURE 3.1 Matrix of generic and site-specific factors.

should have an X placed in the matrix box for that item and should then be considered in the conversations regarding the values generated (minimum, most likely, etc.) for that generic parameter.

For example, at Phil's west-coast facility, the generic factor Site selection would be impacted by Weather, Labor unions, and Long-distance shipment of samples. So, Phil's group would put an X in each of the appropriate boxes on the matrix. Then, when discussing the values to be entered into the minimum, most likely, maximum, peakedness, and probability boxes for Site selection, the group would be instructed to consider the impact of the site-specific items on the values they generate.

Similarly because of labor unions and state-mandated corporate policies regarding the employment of vendors, Phil's group would put an X in the Labor union and Site-specific corporate policies boxes for the generic item Vendor selection. In conversations concerning model-input coefficients for the generic Vendor selection parameter, Phil's group would consider these two items.

With regard to the Vendor selection generic parameter, Phil's group knew that they would have to engage the corporate headquarters legal staff to generate a boilerplate contract. Attached to the contract would be a list of specific tasks to be performed and minimum requirements for vendors who would bid for the contract. Typically, costs associated with generation and distribution of a contract range from a minimum of $5,000 to a most likely of $8,000 to a maximum of $12,000. A probability of 1 (100%) would be used, because group members know that a contract is necessary. A peakedness of 5 might be employed to reflect their uncertainty regarding the actual price of contract generation.

However, Phil's group knew that there were complicating factors at their site. One was the unavoidable interaction with labor unions. Past experience had shown that due to data collection problems and the time required to interact with individual unions, the cost of a "union-free" contract-generation process typically was increased by about 10%. In addition, Phil's group knew that there were state-mandated policies that affected contracts generated at their site. One example was the state mandate that all viable contractors carry sufficient insurance to offset any reasonable liability claims that might arise. It would be up to Phil's group to verify that contractors bidding on the project comply with this mandate. Such verification could take weeks and typically raise the cost of vendor selection by about 15%. Thus, working always from the original base price, Phil's group might employ the following simple calculations, taking into account a 10% increase for the contract-generation process and a 15% raise for insurance verification.

0.1 × $5,000 = $500
0.15 × $5,000 = $750
Minimum Vendor Selection cost = $6,250

0.1 × $8,000 = $800
0.15 × $8000 = $1,200
Most Likely Vendor Selection cost = $10,000

0.1 × $12,000 = $1,200
0.15 × $12,000 = $1,800
Maximum Vendor Selection cost = $15,000

The values of $6,250, $10,000, and $15,000 would be entered on the matrix sheet as the Vendor selection minimum, most likely, and maximum values, respectively. A probability of 1 (100%) would be entered on the matrix sheet for this generic parameter because at Phil's site, unlike some other sites, they know they must hire vendor personnel to aid in the cleanup-assessment stage. A peakedness value of 3 was decided upon to reflect the group's relative uncertainty with regard to the actual cost of this phase of the project. When these data were entered into Don's probabilistic model, a distribution for Vendor selection was generated. That distribution is shown in Figure 3.2.

Don considered Kim's suggestion a brilliant and practical solution to his problem. Upon return to his home site, Don generated a complete matrix that would be

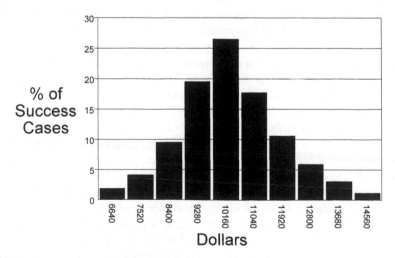

FIGURE 3.2 Frequency plot of total cost for vendor selection.

distributed to all facilities. Don delivered the matrix in person and instructed plant and vendor personnel in the process of selecting site-specific parameters and in the process of considering those parameters when generating coefficients for the generic items. This would take time and effort, but it would accomplish the assigned task in the most time-effective and economical manner.

SO, SO WHAT?

The processes conveyed in this chapter are simple; however, their apparent simplicity belies their power. A long list of advantages is associated with this process. I will mention here but a few.

First, the process described promotes consistency while simultaneously accommodating complexity. Engaging this process enables one to deal with a single risk/uncertainty model and set of input and output parameters, while at the same time addressing the diverse and often arcane aspects of each individual situation. The consistent set of input parameters promotes a common understanding and jargon among those affected by the risk process. The common output parameters facilitate comparison and ranking of assessed projects.

Versatility also is a virtue of this process. Employing the matrix approach allows a single risk model to be used to address a wide variety of situations. This is efficient and cost effective. To accommodate seemingly new situations, there is great tendency among modelers of all types to either build new models or constantly enhance/modify existing models. This approach not only requires considerable expenditures of time and money, but leaves one with either a relatively large set of models or a current model whose output cannot readily be compared to that of a former version of the model. I do not argue here that models should not evolve with time and with changing conditions, but certainly this activity should be kept to a minimum. If models are built using the processes outlined in the first book, revisions of any model should be required less often.

Any risk/uncertainty model is only as good as the input data. When a model is revised often, it is typical that users are inadequately (if at all) prepared to understand or deal with the changes. If a model is relatively stable through time and space, users are better able to understand and provide data for the model input parameters. This greatly enhances the use of archived input and output data. Establishment of a stable process for risk/uncertainty assessment facilitates promotion of the process within an organization and enhances acceptance of the process.

Over the years, I have witnessed the work of many model builders, and it seems that the concept of practicality has been very low on their list of priorities. Many modelers I have known have been most interested in designing and building the "coolest" model possible with as many bells and whistles as could be crammed in. This approach is appropriate when the model is to be executed primarily by a relatively small group of "experts." However, in today's business environment, the number of "experts" employed by corporations is on the decline. The trend is to push work and accountability down to as low a level as is practical. This means that models should be designed with practicality as the prime factor. Using the matrix approach, the number of actual parameters in the model can be kept to a practical number, with complexities of individual situations dealt with via conversations and consensus.

Critics of the matrix approach to risk/uncertainty modeling often point out that this process relies heavily on conversations among participants and, therefore, is less quantitative and rigorous than it should be. I have many a reply to such criticisms. First, no matter how quantitative an analysis might be, if the input data for individual parameters are not determined by data providers engaging in conversations regarding that input data, then I believe that only less-than-optimal input coefficients are likely to result. Unless something (pressures, sales, factory production, etc.) is actually being measured to be input to a model, all coefficients supplied to the model should be discussed and agreed upon by all who have responsibility for the parameters in question. So, conversations regarding models that are maximally quantified should be, in essence, no different from the discussions promoted for the matrix approach.

In addition, I have found through the years that many times the best and most useful part of a risk/uncertainty assessment is the conversation that it fosters. Such conversations certainly facilitate cross-fertilization of ideas, mutual understanding, and in-depth examination of the problem. Often, following such conversations, the stochastic results that will be generated by the risk model are obvious to all, and the process of feeding and executing the model is simply one of documentation and verification.

Those who lean toward a more quantitative approach might also deem the combination-by-conversation approach to melding the effects of subparameters, as used in this chapter's example, to be less than optimal. While using the matrix approach it is typical that participants might discuss in depth the projected effect of labor unions and state-mandated policies on the cost of vendor selection. They might come to agreement regarding the combined impact of these modifying parameters, or they might, as in our example, agree upon a separate impact for each item. In addition, they might express the impact as a percent of a base cost (as indicated above), or they might express the impact in other terms. Similarly, the probability and peakedness factors may also be modified by conversation and consensus.

Those who would take the more rigorous and quantitative approach to the problem might suggest that individual ranges, probabilities, etc., be established for each modifying factor (e.g., labor unions or state-mandated policies) and that these individual ranges be combined mathematically to arrive at a final generic-parameter set of model inputs. Certainly, this can be done; however, I have found it not to be the best approach for several reasons. First, most people are not comfortable generating, or don't have the time to generate, ranges for impacts. Asking them to go through this process once for each generic parameter (to establish a base case) is quite enough. If a generic parameter devolves into multiple subparameters (such as Vendor selection devolving into state-mandated policies and labor unions), it is simply impractical to ask participants to generate reams of probabilistic input for each parameter. In addition, when this approach is taken, the combined impact of the individual subparameters often is never made clear because it is calculated in a computer program. It is often the case that the combined impact is not the impact that the group would intuitively attribute to the situation.

A final point in support of the matrix approach has to do with defending the risk model input and results. While people are engaged in conversation and are being shepherded toward a common understanding of a problem and consensus regarding input parameters, the logic behind the entered data can readily be explained and defended if need be. This is a powerful aspect of the matrix process. Being able to explain the data enhances credibility and acceptance of the model and its output.

There are a few drawbacks to taking the matrix route. First, it obviously takes people to accomplish the task. In many corporations, it is a major accomplishment to actually get all of the necessary parties in one room at the same time (or simultaneously available for a conference call). This aspect of the process should not be underestimated. The best solution I have seen to this problem is for management to deem it part of everyone's job to contribute to such conversations each time an assessment is necessary.

Another difficulty with the matrix approach is that it generally requires a skilled facilitator at each of the meetings. The facilitator needs to be skilled at guiding conversations and keeping control of the room, and must have a relatively complete understanding of the model and probabilistic approaches in general. Such individuals can be tough to find.

4 New Manufacturing Facility — Business-Justification Model

CONTENTS

INTRODUCTION

When considering the construction of a new manufacturing facility, most companies are concerned with two primary aspects of the project. First, engineers are required to plan the physical facility. Site selection, physical size, equipment, and required raw materials are just a few of the considerations addressed by engineers.

Equally important is a business analysis for the new facility. Corporate executives must know whether the new facility will generate the financial returns needed. Economists usually are called upon to plan and execute the business analysis. In the hypothetical example below, we will follow an economist through the requisite steps of such a study.

WHAT IS THE QUESTION?

Shelah is the chief economist for BarMade (BM), a small specialty-steel-products manufacturing firm. Although BM does ship a small portion of its products to nonadjacent states and export to some overseas locations, most of its products are consumed in a tri-state area.

Recently, a nearby large municipality announced a construction project estimated to last a decade or more. The proposal is to build a mass transit system to connect suburban locations to the downtown business hub. As part of the project, special steel reinforcing rods will be needed for track construction. City officials and engineers have had preliminary talks with BM management regarding the feasibility of their manufacturing such special rods in large quantities over the span of the project. BM management believe that a new manufacturing facility might have to be built to handle the production of these special rods.

Company executives have assigned chief economist Shelah to spearhead the probabilistic business analysis. There are many economic, political, technical, logistical, and other parameters to consider. However, in a typical management-type move, Shelah has been given the responsibility to effect the analysis, but no increase in authority to get the job done. Successful execution of the probabilistic analysis will depend heavily upon the cooperation of various segments of the company over which she has no direct authority. In addition, it is unclear to Shelah just what question she is supposed to attempt to answer.

It is apparent to Shelah that the proposed new plant will initially be dedicated to production of the special rods needed for the mass transit project. Production of the rods, she knows, is the sole focus of the engineers responsible for the design of the new plant and procuring equipment. Shelah has to consider whether a short-term, single-product venture like the one proposed can justify the commitment of funds of the magnitude necessary to build the new facility.

Can the plant design accommodate different equipment after the conclusion of the mass transit project? Can a new purpose be found for the facility? Even if a new purpose can be established after the conclusion of the proposed mass transit project, can the facility be retooled and re-equipped economically? Should the strategies and capabilities of present and future competitors be considered in the analysis, or should the justification for the project be predicated on the assumption that BM will be the selected and sole provider of the special parts? Can reasonable cost and price projections be made for a special product that does not yet exist? These and other questions are just a few of the serious problems Shelah must resolve prior to the design of her economic risk model.

Shelah will leave it to the company metallurgists to determine whether it is possible and practical to produce a steel that will meet the demands of the mass transit project. She will need to know, however, the composition of the new steel so that reasonable manufacturing costs and prices can be established. Small changes in composition can dramatically affect the cost of product production. Even a small amount of titanium in the steel composition, for example, can drastically increase the price of the finished product.

Shelah has had years of experience generating economic analyses for production of conventional steels and the plants that produce them. However, the steel to be manufactured for this project will not be conventional in nature. Unconventional steel might require unconventional equipment to produce it. Shelah realizes that she will need to understand the type and cost of potentially special equipment. This will require almost immediate meetings with design engineers who themselves are likely to be reeling from the pressure and have many questions of their own. It will be a monumental task to get them to agree to a meeting, much less glean useful information from them.

One bright spot in an otherwise mottled landscape is that Clive, the head design engineer, is an old friend of Shelah's with whom she has collaborated numerous times in the past. Shelah is desperate to define the economic problem to be resolved, but realizes that a meeting with Clive is a prerequisite to any attempt at problem delineation.

Clive's dilemma is similar to Shelah's. Before he can design a physical facility, he must learn from the metallurgists just what type of steel must be produced. Clive informs Shelah that he will attempt to obtain preliminary compositional data from the metallurgical team and will respond to her request for a meeting upon receipt of the information.

Company metallurgists have just concluded the first of what will prove to be many meetings with Transit Authority engineers. The purpose of the meeting and of future meetings is to establish the job to be performed by the BM-manufactured parts. From their initial meeting, it is clear to BM metallurgists that the rods provided will have to retain rigidity in a relatively high-temperature environment. To meet such performance standards, metallurgists deduce that the composition of the special steel will require incorporation of unconventional components. The first-guess composition of the steel is generated and related to Clive. Clive, in turn, conveys this information to Shelah in their initial meeting. Shelah realizes that the metallurgists will go through multiple compositional iterations prior to establishing a final formula for the steel, but the estimated composition is the formula upon which she will have to base her economic estimates. Using distributions in her probabilistic analysis she is confident that she can account for the uncertainty associated with questions concerning the steel's composition.

Clive also relates to Shelah that he expects production of the special steel will require use of unconventional ovens and curing processes. Although he is unsure about the amount or cost of such equipment, he knows that Shelah will require his best guess regarding the physical plant. In his meeting with Shelah, he relates his preliminary estimates.

Shelah understands that she is exceedingly fortunate to have even preliminary estimates for steel composition and physical-plant design so early in the game. Her personal relationship with her coworkers is the sole reason for the expedience and cooperation. Shelah also realizes that she has received just about all the help she is likely to get in the preparation of her economic model. Other decisions she will have to make on her own.

Shelah knows that if she queries company management regarding any post-mass-transit-project use of the proposed facility, management response will be boilerplate and noncommittal. She decides, therefore, that economic justification for the new facility will consider only the mass transit project. This decision, in turn, establishes a time frame for her economic model. The mass-transit project is to be fundamentally completed in 10 years. Shelah's model, therefore, will be a probabilistic time-series model that considers project economics over a decade.

Although she will ignore any post-mass-transit-project use of the facility in her analysis, Shelah concludes that she cannot ignore the economic ramifications posed by competing specialty-steel makers. While BM, because of reputation and logistics, is the logical provider for the special steel rods, the transit project is lucrative enough to attract at least one well-known competitor. Award of the contract, product price, and plant utilization are just three of many parameters that might be adjusted because of competitive considerations.

Armed with information provided by Clive and her own conclusions, Shelah convenes a meeting with her staff. After much discussion, the team concludes that the economic model should attempt to answer the question: "Considering a 10-year time frame, posturing of competitors, and exclusive use of the facility for the manufacture of transit-project rods, what is the NPV (net present value) to the company of the construction of the new facility and sale of resulting products?"

CONSTRUCTION OF THE CONTRIBUTING-FACTOR DIAGRAM

Shelah knows that time is of the essence. Contracts will be let for the new project within a month. The typical meticulous, cumbersome, spreadsheet-analysis approach is not an option for determining whether BM should bid on the project. Certainly spreadsheet-based conventional approaches will be appropriate during the project, if and when BM lands the contract. For the task at hand, however, a more adroit, flexible, and "high level" probabilistic risk model will be employed.

In her first working meeting with her team, Shelah realizes that balancing "high level" with meticulous tendencies and detail will pose the greatest challenge. Shelah also perceives that she is too close to the problem to impartially conduct a design session for the risk model. She engages the services of Howard, a skilled risk-model design expert and a gifted facilitator.

Howard tells the accounting group that he is impressed that the group has successfully defined the question to be answered and says that the next step is to define a few categories for the risk model. These categories will then be populated with variables. The variable collection will be designed to be comprehensive yet high level — a tricky proposition.

Howard knows few specifics about the production of specialty steels. He is, however, quite astute at designing general business risk models. He states that, fundamentally, each business model has two basic parts — income considerations and expenses. The group agrees and settles on the names "Volume/Price" for the income category and "Costs/Rates" for the expense category. Shelah points out that there are other considerations. She is concerned that, although the municipality is "all fired up" regarding the transit project, the voting population has yet to agree to bond issues which will provide partial funding for the project. The timing of such a thumbs-down might come after considerable capital outlay by BM. Shelah also would like to include a parameter reflecting the chance that BM will gear-up for the project, but that BM's primary competitor will be awarded the lion's share or all of the work. Howard agrees, and suggests the inclusion of a third category to accommodate such chances of abject failure.

The next task is to populate the categories with pertinent variables. It is quickly agreed that three fundamental aspects of the Volume/Price category are capacity of the new plant, utilization of that capacity, and the price for the new product. In addition, Shelah suggests the inclusion of two parameters that will accommodate reflection of the influence of a competitor on price and utilization. It is agreed that these concerns should be included. Ramifications for capacity, relative to competitor actions, are considered moot because BM must commit to a physical plant design long before it learns what the competition might do.

The risk model will be a time-series probabilistic model, which will consider the life of the plant over a ten-year period. Such consideration requires that data for at least some input parameters be gathered for each of the ten years. The group agrees that it is impractical, and likely impossible, to require much detailed information for all model parameters. Time and extra-departmental cooperation preclude it. Therefore, the team decides to collect, in the Volume/Price category, yearly data on only plant utilization and price. To expand the single-year deterministic estimates for utilization and price into distributions, the team agrees to include two expansion distributions in the category.

Team members agree that the Costs/Rates category should include only basic variables that reflect fixed operating costs, capital expenses, variable costs, incentives, the currently used corporate discount rate, and the corporate tax rate. As with the income-related variables, team members declare that they will need yearly data and associated expansion factors for only two of the variables — operating costs and capital expenses.

Howard, Shelah, and the team also conclude that there are only two non-financial aspects that might cause BM's involvement in the project to fail outright: (1) After BM has realized considerable project-related investment, voters will deny critical funding for the project, effectively killing it, and (2) BM's primary competitor will land the contract so that BM is completely excluded or excluded to a significant extent. The contributing-factor diagram resulting from the meetings is shown in Figure 4.1.

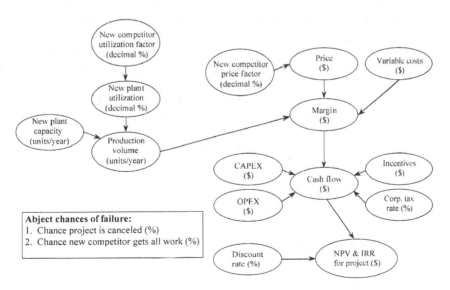

FIGURE 4.1 Contributing factor plot for new manufacturing facility — business-justification model.

TABLE 4.1
Distribution-Based Variables

Variable	Minimum	Most Likely	Maximum	Peakedness
New Plant Capacity (Units)	1,000,000	5,000,000	8000,000	3
New Plant Utilization Expansion Factor 2001-2005 (Decimal %)	0.75	1	1.1	5
New Plant Utilization Expansion Factor 2006-2010 (Decimal %)	0.95	1.2	1.5	5
New Competition Utilization Factor (Decimal %)	-0.3	-0.2	-0.1	10
Price Expansion Factor (Decimal %)	0.9	1	1.3	5
New Competition Price Factor (Decimal %)	0.7	0.8	0.9	10
Fixed Operating Cost Expansion Factor (Decimal %)	0.9	1	1.2	5
Capital Cost Expansion Factor (Decimal %)	0.9	1	1.3	5
Variable Costs ($)	500,000	600,000	800,000	5
Incentives ($)	500,000	600,000	700,000	5
Discount Rate (Percent)		11		
CorDorate Tax Rate (Decimal %)		0.5		

POPULATING THE MODEL WITH DATA

Howard next guides the group through several days of meetings in which the group is challenged to generate data for the critical input parameters. Using the data provided by Clive and their own business acumen, the group generates the data that appear in Tables 4.1 through 4.3.

Plant capacity is a major concern. The group's best estimate is that BM will need to produce about 5 million units per year in initial years. It is not clear, however, just how many rail spurs (lines effectively built in series) will be started in the first year. If only one spur is started, then perhaps only 1 million units per year will be required. If, however, all rail lines are initiated in the first year (effectively built in parallel) then as many as 8 million rods might need to be produced annually. A peakedness of 3 is assigned to reflect the uncertainty in the capacity estimate. The resulting capacity distribution is shown in Figure 4.2.

TABLE 4.2
Chance of Failure Variables

	Chance of Failure
Chance project will be canceled	5
Chance new competition gets all work	15

TABLE 4.3
Time-Series Variables

Year	Utilization (Decimal %)	Price ($/Unit)	Fixed OPEX ($)	CAPEX ($)
2001	0.65	2	1,000,000	20,000,000
2002	0.66	2.1	1,000,000	1,000,000
2003	0.67	2.2	1,000,000	1,000,000
2004	0.7	2.3	1,000,000	100,000
2005	0.73	2.5	1,000,000	100,000

Regardless of whether lines are initiated in a series or parallel fashion, it is typical of such projects for demand for products to build to a production crescendo near the conclusion of the project. The team concludes that plant utilization will initially be about 65%, building to an average of about 80% in the final years of the project. The unexpanded utilization time series is shown in Figure 4.3.

Under the assumption that BM will be awarded the contract and reflecting the group's unanimous feeling that it is better able to predict near-term events relative to those in the more distant future, the group agrees to generate two expansion distributions for plant utilization. The first expansion distribution addresses uncertainty in the first 5 years of the project. The group believes that its yearly utilization estimates for the initial half-decade might range from as much as 25% too high to 10% too low. In nearly every publicly funded project, contractors generally run behind their optimistic preproject estimates. Therefore, final 5-year estimates should reflect that the utilization values might only be 5% too high but could be overly conservative by up to 50%.

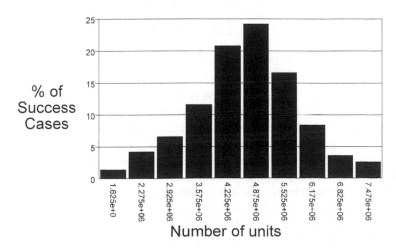

FIGURE 4.2 Plot of new plant capacity.

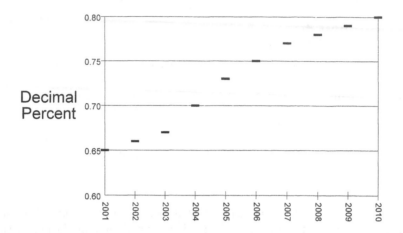

FIGURE 4.3 Time series of new plant utilization.

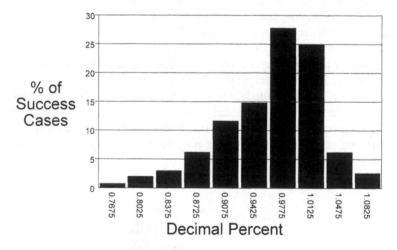

FIGURE 4.4 New plant utilization expansion factor for the years 2001 to 2005.

Competitor effect on plant utilization is also considered. The team believes that actions of competitors could reduce utilization from between 10% to 30%. The two expansion distributions, the competitor-effect distribution, and the expanded utilization time-series plot are shown in Figures 4.4 through 4.7, respectively.

Based upon Clive's steel-composition data and other specifications, Harold, Shelah, and the team decide that rods could initially be produced for about $2 per unit. Clive, however, admits some conservatism in his estimates. In addition, engineers reporting to Clive have indicated that they believe the initial specifications for the rods supplied by the municipality engineers are too conservative. Clive's engineers believe that the rods ultimately will need to be longer, thicker, and able to withstand higher temperatures than indicated by the municipal engineers. If Clive's

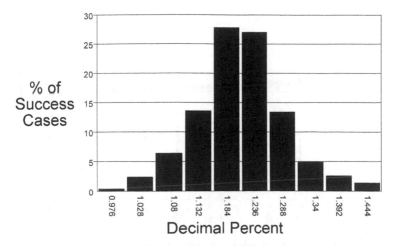

FIGURE 4.5 New plant utilization expansion factor for the years 2006 to 2010.

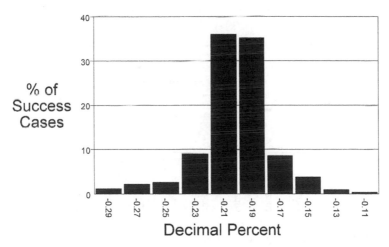

FIGURE 4.6 New competitor utilization factor.

engineers are correct, the cost of the rods will increase with time. Considering all these factors, the team estimates that the price of the rods could be as much as $3.5 per unit in the final year of the project. The unexpanded time-series plot of per-unit price is shown in Figure 4.8. The expansion factor for price shown in Figure 4.9 indicates that the team feels that its estimates for price could run from 10% too high to 30% too low.

Competitor actions also are expected to influence price. Given its knowledge of its competitor's costs and price structure, the team estimates that competitor action could cause BM to reduce its estimated yearly prices by 10% to 30%. The competitor price-factor distribution and the expanded time-series plot are shown in Figures 4.10 and 4.11, respectively.

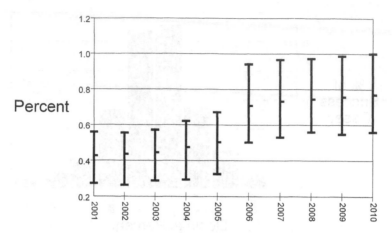

FIGURE 4.7 Time series of expanded new plant utilization.

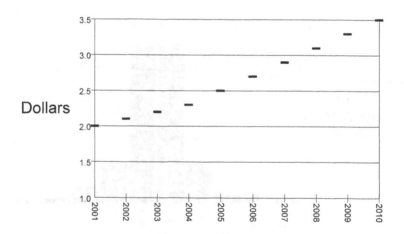

FIGURE 4.8 Time series of per-unit price.

On the cost side of the equation, the team concludes that there are just too many unknowns to be able to make reasonable operating-cost estimates for this particular plant. Although BM might need to purchase special equipment to manufacture the rods, operating costs for the equipment are not expected to be widely divergent from those for conventional machinery. It is also expected that manpower costs will be typical. For other conventional plants, operating costs have averaged about $1 million per year. The team decides to use the $1 million figure for all years in the preliminary operating-cost time series. The associated expansion distribution indicates that team members believe that the $1 million figure could be from 10% too high to 30% too low. The expansion distribution is shown in Figure 4.12.

The major capital expenses associated with plant construction are expected to be absorbed in the first year. All purchases for non-perishable raw materials and

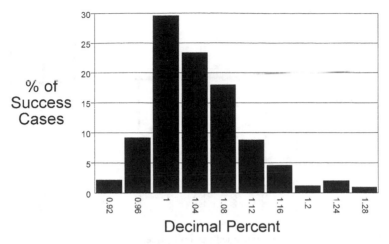

FIGURE 4.9 Expansion factor for per-unit price.

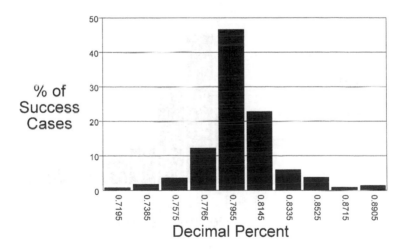

FIGURE 4.10 Competitor price factor distribution.

equipment will be made in year one. First-year capital costs are expected to be approximately $20 million. Capital expenses are expected to drop to about $100,000 per year through the fifth year of operation. Increased production of redesigned rods and a push to meet deadlines are expected to increase capital spending to approximately $200,000 in the final year. The associated expansion distribution for capital spending indicates that the team believes their initial estimates could be high by about 10% or low by as much as 30%. Plots of expansion distribution and the expanded capital-cost time series are shown in Figures 4.13 and 4.14, respectively.

Yearly variable costs for the project are expected to range from $500,000 to $800,000. Incentives associated with the project are envisioned to range from about

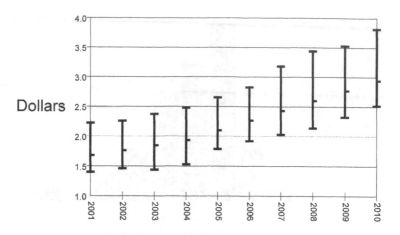

FIGURE 4.11   Time series of expanded per-unit price.

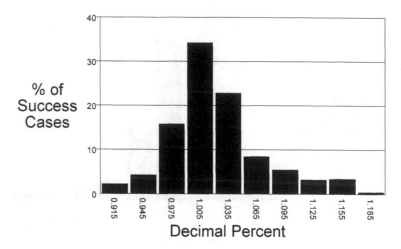

FIGURE 4.12   Expansion distribution for fixed operating costs.

$500,000 to about $700,000. The current discount rate is 11% and the corporate tax rate is 50%.

Chances of abject failure are those associated with cancellation of the project by voters and with awarding the contract to a competitor. Public opinion polls seem to indicate favorable public reception of the mass-transit idea. Team members, therefore, believe that there is only about a 5% chance that the project will be killed by the voters.

Award of the contract to a competitor is perceived as a more likely cause of abject failure. Although BM is the logical choice with respect to logistical proximity and technical prowess, the competing company has several other product lines and might offer to produce the rods at or below cost. This is not a likely scenario, but

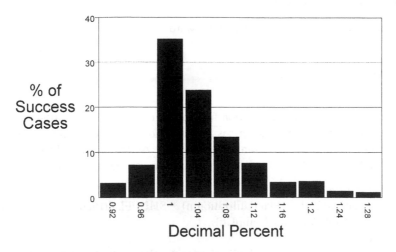

FIGURE 4.13 Expansion distribution for capital costs.

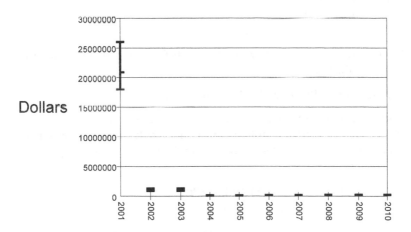

FIGURE 4.14 Time series of expanded capital xepenses.

it cannot be discounted. Team members believe there is about a 15% chance that BM will lose the contract to the competition. The total chance of abject failure for BM is calculated below.

Total Chance of Failure (TCF) =
$$1 - (1 - 0.05) \times (1 - 0.15) =$$
$$1 - (.95 \times .85) =$$
$$1 - 0.8075 =$$
$$0.1925 \times 100 = 19.25\%$$

RISK MODEL EQUATIONS

The risk model is relatively straightforward. Computer code for the model is shown below.

```
n = 0;
while n < _NUM_PER do
```

"Utilization is equal to utilization for that year times the utilization
"expansion factor which generates a distribution from the single-values
"utilization number entered in the time-series. Added to the utilization is
"the reduction (value is negative) in utilization caused by new competitors.

```
  if n < 5 then
    PlantUtil2[n] = (PlantUtilization[n] * SHUFFLE(PlantUtilExp0105)) +
        SHUFFLE(NewCompUtil);
  else
    PlantUtil2[n] = (PlantUtilization[n] * SHUFFLE(PlantUtilExp0610)) +
        SHUFFLE(NewCompUtil);
  endif
  if PlantUtil2[n] > 1.0 then
    PlantUtil2[n] = 1.0;
  endif
```

"Volume for the year is equal to utilization for that year multiplied by the
"capacity for that year.

```
  Volume[n] = PlantUtil2[n] * PlantCap[n];
```

"Price is equal to the entered price times the price expansion factor
"which creates a price distribution from the single-valued prices entered
"in the time-series variable. This is multiplied by the new-competitor
"price factor.

```
  Price2[n] = (Price[n] * SHUFFLE(PriceExp)) *
      SHUFFLE(NewCompPrice);
```

"Contribution margin for the year is equal to the volume multiplied by the price
"minus the variable costs.

```
  ContMargin[n] = (Volume[n] * Price2[n]) - SHUFFLE(VarCosts);
```

"Cash flow is equal to the contribution margin for that year minus the fixed
"operating costs for that year times the fixed-operating cost expansion
"factor plus the capex times the capex expansion factor plus the incentives.

```
  CashFlow[n] = ContMargin[n] - ((FixedOpex[n] *
      SHUFFLE(FixedOpexExp)) + (Capex[n] * SHUFFLE(CapexExp))) +
      SHUFFLE(Incentives);
  n = n + 1;
enddo
```

NPV = NPVSIMPLE(CashFlow*(1 − CorpTax),DiscRate);
IRR = IRRSIMPLE(CashFlow*(1 − CorpTax),DiscRate);

where

n is an internal counter

_NUM_PER is a system variable that holds the number of periods (10 in this case)

PlantUtil2 is the plant utilization considering the utilization expansion factor and the effect of competition

PlantUtilization is a time-series of deterministic yearly plant-utilization values

SHUFFLE is a system variable that causes a distribution to be re-sampled across time periods within an iteration

PlantUtilExp0105 is the utilization expansion distribution for years 2001 to 2005

NewCompUtil is the estimated impact on utilization from competitor

PlantUtilExp0610 is the utilization expansion distribution for years 2006 to 2010

Volume is a time-series output variable that represents yearly plant utilization multiplied by the plant capacity

PlantCap is the estimated new plant capacity

Price2 is a time-series output variable that holds the expanded yearly price multiplied by the effect of the competition

Price is a time-series of deterministic per-unit prices for each year

PriceExp is the price expansion distribution

NewCompPrice is the distribution representing the effect on price from competitors

ContMargin is a time-series variable that holds values for the calculated contribution margin

VarCosts is the distribution of estimated variable costs

CashFlow is a time-series output variable that holds values for the calculated project cash flow

FixedOpex is the time-series of deterministic estimates of yearly fixed operating expenses

FixedOpexExp is the fixed operating expenses expansion distribution

Capex is the capital expenses for a given year

CapexExp is the capital expenses expansion factor

Incentives is the distribution of estimated incentives

NPV is the net present value of the project

NPVSIMPLE is a system variable that automatically calculates an NPV from a series of cash flows

CorpTax is the corporate tax rate

DiscRate is the corporate discount rate

IRR is the internal rate of return for the project

IRRSIMPLE is a system variable that automatically calculates an IRR from a series of cash returns

RESULTS FROM MODEL EXECUTION

Inspection of the Plant Utilization plot (Figure 4.7) shows that maximum-utilization rates in the first 5 years of the project will generally be below 60%. In the final 5 years, utilization rates will most likely be near 75% but could approach 100% (see 2010 frequency plot of utilization in Figure 4.15).

Figure 4.16 of Plant Production Volume indicates that median production will range from about 2 million units in initial years to a maximum of about 3.6 million units in later years. Peak production in later years could reach 7 million units, but, as shown in Figure 4.17, which depicts the production distribution for year 2010, this is not likely, Median prices are expected to climb from a first-year low of about $1.7 per. unit to a final-year high of about $2.9 per unit as shown in Figure 4.18.

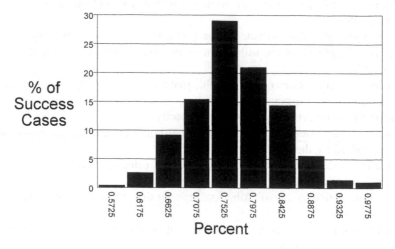

FIGURE 4.15   Time series of plant utilization in the year 2010.

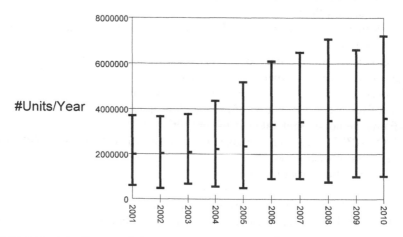

FIGURE 4.16   Distribution of plant utilization in the year 2010.

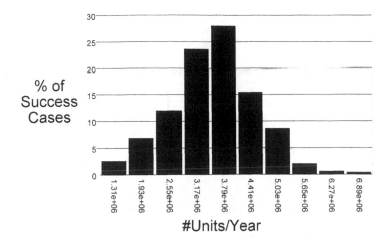

% of
Success
Cases

#Units/Year

FIGURE 4.17   Distribution of plant production volume in the year 2010.

A steeper increase in price is evident in the final 5-year period. Median margins, as shown in Figure 4.19, are expected to be as low as $2.5 million in the first year. Median margins will increase to approximately $10 million in the final year of the project.

Cash flow for the project is shown in Figure 4.20. Second-year cash flow has about a 5% chance of being negative in addition to the almost 20% chance of abject project failure. The net risk-weighted cash flow for the second year is expected to be about $1.2 million. Final-year cash flow shows no negative downside. The net risk-weighted cash flow in year 10 is about $7.5 million (Figure 4.21).

Inspection of the project NPV plot (Figure 4.22) shows that there is about a 65% chance that the project will yield a positive NPV. There is a 50% chance that the project will yield an NPV of about $3 million or more. The net risk-weighted NPV for the project is about $3.1 million. From Figure 4.23 Shelah can see that the net risk-weighted IRR is about 18%. Shelah will carry these values forward to management for portfolio management purposes and for determining whether the returns for the project are sufficient to justify construction of the new plant.

SO, SO WHAT?

I particularly enjoyed creating this example because it reflects some of the actual work-a-day-world realities to which all of us are subject. When in need of critical information or a service, the bottom line is personal relationships matter. Corporations often do not recognize, or at least are unwilling to acknowledge, that personal relationships among employees are a primary means by which business gets done. "Unofficial" networks are critical to the successful execution of business.

On a more technical note, this chapter demonstrates the ability of the Monte Carlo process to deal not only with risks, but with simple uncertainties as well. For example, in this example there was no risk that there would be no composition for

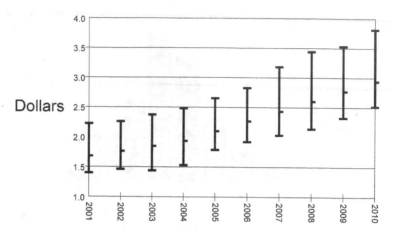

FIGURE 4.18   Time series of price.

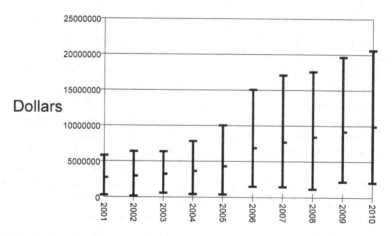

FIGURE 4.19   Time series of margins.

the steel; however, there was considerable uncertainty surrounding the issue of exactly what the composition and, therefore, the price might be. Similarly, there was no intended risk of whether a spur would be built, but there was uncertainty surrounding the issue of just how many spurs would simultaneously be attempted. The Monte Carlo method facilitates the integration of all uncertainties to effect the output variables. This is one of the salient benefits of the process.

Model validation is always an issue. As stated in the Introduction of this book, in the real world it is often impractical to hope for validation of a risk/uncertainty model's output. That is, if a decision must be made, it can be made with or without the use of a risk/uncertainty model. If this model is used, then the decision will be taken and the consequences realized. If the consequences are fair or good, you are

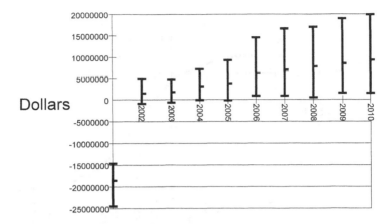

FIGURE 4.20 Time series of cash flows.

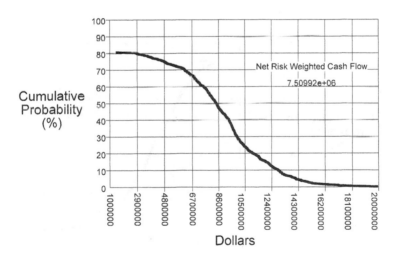

FIGURE 4.21 Cash flow distribution in year 2010.

happy and might determine that you are glad you did not go through the time-consuming process of generating a model. If the results are poor, however, you might wonder whether enlisting a model would have helped. The same logic applies if you initially decide to use a model. If things turn out well, you are pleased and give some credit to the model approach, but if things turn out poorly, you have to wonder whether you would have done better without all the bother of the model.

The point is, rarely in real life are we allowed to compare one scenario with another. We are forced to decide "up front" which way we will go and then are committed to that path. We generally are not allowed to know the results of the

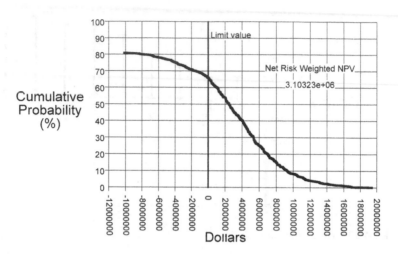

FIGURE 4.22   Plot of net present value.

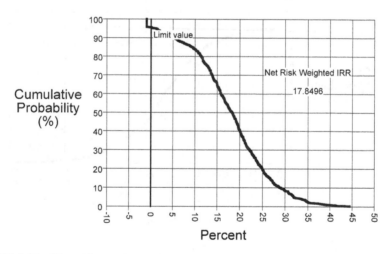

FIGURE 4.23   Plot of internal rate of return.

"road not taken." You, of course, can compare the just-realized result with historical results, but this must be done carefully — both practically and statistically.

Validation of any approach is best measured in terms of satisfaction. If you apply a model and are satisfied with the result, it is not much more than an exercise in torture to attempt to determine just how the realized result would stack up against a result from a different process. If, over time, results from multiple application of the model approach are generally satisfactory, then validation of the model and surrounding process must be acknowledged.

Focusing on only the results of a risk/uncertainty model considers but a small part of the overall benefit of such an approach. Some of the greatest enhancements to a business come not from the technical aspects of this process, but from the social and organizational facets. For example, the discussions regarding the ranges of values for parameters are often much more useful than the actual values that result from such discussions.

This project was purposely constructed so that the results would be ambiguous. That is, when Shelah finished her analysis, the resulting financial measures neither completely condemned the project nor indicated that it was a shining business opportunity. The risk/uncertainty-model approach does not (or should not) make decisions for you. Just because you have gone through the trouble of generating and populating a model does not mean that you will not have to use judgment to finally decide what to do. Given the results of her analysis, if you were Shelah's management, what would you decide to do?

5 Oil-Field-Development Investment — Opportunity Risk Model

CONTENTS

INTRODUCTION

Part of the job of an investment broker is to ferret out worthwhile opportunities in which a bank or individuals can invest. Entrepreneurial or small independent oil-field-development companies typically are engaged in never-ending quests to attract such investors. Such companies, after successfully funding and drilling an exploration well, have to seek funding to develop the oil field. In the scenario described below, one such development scheme will be considered.

Ivan is an investment banker at Barney & Ruble (B&R), an eastern European financial institution. Ivan handles investments in the energy sector. When times were good for the energy business, Ivan typically received about one solicitation per month from small energy firms seeking financial backing for a venture. With the current glut of oil on the world market and the subsequent downturn in commodity prices, the number of small firms seeking financial aid has dramatically increased.

In the past, Ivan has used personal contacts and visits to gather data on prospective clients. When he had received a request for funding, his first move was to reach for the telephone and speak with the party requesting funds. Typically, the phone conversation would result in a face-to-face meeting with potential clients. Sometimes clients traveled to Ivan's office for such meetings. More often, Ivan traveled to the requestor's site to gather critical data firsthand and form an opinion about the viability of the potential client's business.

Bad times in the "oil patch" have caused Ivan to rethink his investigative and approval processes. He now receives from 5 to 20 requests per month for financial aid. Given the magnitude of the typical investment, B&R can afford to fund only 3 to 5 requests per month. In the past, Ivan's "visit the customer" practice required approximately 7 to 14 days. It is clear to him that the old process is too lengthy and cumbersome to meet the current demands. He must try something new.

Ivan has some background in operations research and turns for help to a former classmate who works in the field. The classmate quickly deduces just what Ivan needs and puts him in contact with Risk and Decision Making, Inc. (RDM), a western firm that consults with and builds risk models for companies worldwide.

WHAT IS THE QUESTON?

Glenn is the RDM representative who handles the energy sector. From their initial meeting, it is clear to Glenn that Ivan needs help, but that Ivan is confused as to just what he needs.

It is evident that Ivan would like to be able to process a greater number of applications in less time. Much of Ivan's decision-making process in the past was based upon the personal opinions he formed after visiting potential clients. Given that the degree of exposure to a potential client's site must now be severely curtailed or nonexistent, Glenn points out to Ivan that his decision making will need to be based on other criteria. One of Glenn's first questions to Ivan concerns just what criteria Ivan would like to use to assess, compare, and rank fund seekers. At first, this question seems trite to Ivan. However, when Glenn requests that Ivan actually list on a piece of paper no more than five items that might be used in customer evaluations, Ivan hits a conceptual wall.

Ivan's fundamental intuitive- and personal-experience-based process of the past incorporated many tangible and intangible parameters. When Ivan made a visit, he typically, consciously or subconsciously, formed an opinion of the person with whom he met. In addition, Ivan evaluated the physical facility, judged the current amount of activity, formed an opinion regarding the type and quality of the equipment to be employed, and performed many other evaluations. Glenn points out to Ivan that these sorts of measurements, for the most part, will no longer be practical and that any new evaluation process will need to be based upon easily obtained and mainly tangible criteria.

Glenn has much experience in performing such evaluations. He realizes that if Ivan is to process up to 20 evaluations per month, any system used to enact such investigations must be practical to run, contain only the most critical information, and require from the customer only information that can be supplied readily.

To identify what problem is to be solved first, Glenn requests from Ivan a list of the types of projects submitted over the past 12 months. Table 5.1 shows the projects and categories.

Because field-development projects were the dominant project type, Glenn submits that the initial risk model address field-development evaluations. The next question is the difficult one. What criteria does Ivan wish to use to compare and rank the opportunities?

Although Ivan has in the past gathered a significant amount of intangible data upon which he based his recommendations to his superiors, the fund-granting individuals at B&R used typical business measures as the basis of their decisions. Upon querying these individuals, Ivan and Glenn learn that when Ivan recommended a

---

TABLE 5.1
Project Number and Type

Number of Projects	Project Type
19	Oil/gas field development
10	Exploration wells
11	Acquisitions

---

potential customer to B&R management, the bosses would require the B&R economists to gather pertinent data regarding the customer and generate a cash-flow and net present value (NPV) analysis. Glenn suggests, then, that Ivan's model should result in an NPV plot and require as input only relevant data. Ivan agrees.

CATEGORIES AND VARIABLES

Because they now know what the target for the risk model is, Glenn suggests that only a few high-level categories be established for the model. After some discussion, they both agree that a simple high-level model would include only fundamental categories for technical considerations, revenue, costs, and chances of failure.

The technical category will contain variables that relate to the physical attributes of the field to be developed. Only a few points need to be considered.

Ivan indicates that one fundamental variable is the average size of economic fields (multi-well drainage area). Another factor is the number of such drainage areas that will be successful in the development project. The number of successful fields multiplied by the average size of the fields should generate a plot that will indicate the size of the resource (oil or gas) base represented by the project.

Ivan knows that there is another complicating factor. These projects take time. All field-development wells cannot be drilled (nor should they be) in one year. To account for this, Ivan submits that the model should provide a means for indicating the percent of the successful fields that will be developed in each year of the project.

Both Glenn and Ivan know that only a portion of the oil in place in a field can effectively be recovered. Therefore, it is deemed pertinent to include a variable to represent the percent of the oil in the underground reservoir that can economically be brought to the surface.

The revenue category is next addressed. Both men agree that for each barrel of oil brought to the surface, a sales price needs to be assigned. They agree to, at first, assign prices on a yearly basis. The price in any given year will represent their estimate of that year's average price for a barrel of oil.

Ivan introduces one more factor. He points out that wells in the area produce significant amounts of sulfur as a by-product, and there are industries in the area to which the sulfur can be sold. Ivan suggests, therefore, that a variable be introduced to account for this incremental value chain. Glenn agrees, and the variable is added to the model. Also included in the revenue category are two "boilerplate" variables for the required rate of return for the corporation and for the current tax rate.

The cost category will contain variables that account for three major costs. Ivan and Glenn agree that the deal-making/breaking costs are those associated with the purchase of equipment and other infrastructure items, the costs associated with geologic and geophysical studies and surveys, and personnel costs related to all field-development and operations activities.

Next to be considered are the factors that could cause the field-development project to fail abjectly. Both Ivan and Glenn realize that such projects are fraught with perils.

One of the most typical sources of failure for such endeavors is the chance that the geologic/geophysical interpretation upon which the project is based is wrong. Such interpretations are about equal parts art and science. There are many complicating factors over which interpreters have no control and about which assumptions must be made. Miscalculations involving trend size, trap closure, rock type, and other parameters are easily made. Any single misinterpretation or combination thereof can torpedo the entire project.

Technical considerations also present a risk of abject failure. For example, there is a chance that drillers will encounter subsurface conditions for which they do not have the technology to handle. Highly over-pressurized zones are a typical example.

Other nontechnical risks also exist. For example, there is some chance that the project will be canceled if market conditions become unacceptable. Low demand for the product and a precipitous drop in oil prices are two strong possibilities. In addition, because the contracts between all parties have not been finalized and signed, there exists some chance that unacceptable contract terms could end the project.

Political and environmental threats must also be considered. Political restrictions such as bans on drilling, zoning restrictions, and other politically related problems could kill the effort. Likewise, environmental constraints and restrictions can make any project unattainable. To capture and validate the agreed-upon set of variables and logic train, Glenn generates a contributing-factor diagram. The result of his effort is shown in Figure 5.1.

FIELD-DEVELOPMENT RISK MODEL EQUATIONS

Equations for this high-level model are simple.

```
n = 0;
while n < _NUM_PER do
    cost[n] = (infrastructure[n] * SAMPLE(infra_var,1)) + (gg[n] *
        SAMPLE(gg_var,1)) + (people[n] * SAMPLE(people_var,1));
    if n = 0 then
        total_cost = cost[0];
    elseif
        n > 0 then
        total_cost = total_cost + cost[n – 1];
    endif
    undiscovered = field_size * 1000000 * no_fields * recovery_factor;
    undisc_year[n] = undiscovered * pct_dis_peryr[n];
```

```
    revenue[n] = undisc_year[n] * (pv_boe[n] * SAMPLE(pv_boe_var,1)) +
        (value_chain[n] * SAMPLE(value_chain_var,1));
    cash_flow[n] = revenue[n] – cost[n];
    n = n + 1;
enddo
NPV = NPVSIMPLE(cash_flow * (1-tax_rate),ror);
```

where

n is a variable used for counting time periods

_NUM_PER is a system variable that contains the number of time periods (years, in this case) over which the model will run

cost is a time-series variable that holds the yearly combined capital and operating costs

infrastructure is a time-series variable that, for each year, holds the costs associated with equipment purchases and development of infrastructure

SAMPLE is a system variable that causes resampling of a distribution in each time period within an iteration

infra_var is a variable that holds a distribution that is used to expand single-valued time-series coefficients into distributions

gg is a time-series variable that, for each year, holds the cost associated with geological and geophysical activities

gg_var is a variable that holds a distribution that is used to expand single-valued time-series coefficients into distributions

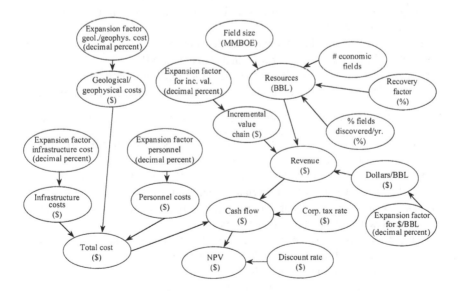

FIGURE 5.1 Contributing-factor diagram for the oil-field-development investment-opportunity risk model.

people is a time-series variable that, for each year, holds the cost associated with the cost of personnel

people_var is a variable that holds a distribution that is used to expand single-valued time-series coefficients into distributions

total_cost is an output variable that holds the distribution of total costs for the project

undiscovered is an output variable that represents the distribution of total undiscovered resources in all the fields

field_size is a distribution of the average field size

no_fields is a distribution of the number of fields that exist in the play trend, i.e., the series of geographically associated fields

recovery_factor is a distribution of recovery factors

undisc_year is a time-series variable that holds the undiscovered resources for each year

pct_dis_peryr is a time-series output variable that, for each year, holds the percent of fields discovered in that year

revenue is a time-series output variable that holds, for each year, the distribution of revenue generated by the project

pv_boe is a time-series variable that, for each year, holds the average yearly price for oil

pv_boe_var is a variable that holds a distribution that is used to expand single-valued time-series coefficients into distributions

value_chain is a time-series variable that, for each year, holds the average yearly value from sale of secondary products

value_chain_var is a variable that holds a distribution that is used to expand single-valued time-series coefficients into distributions

cash_flow is a time-series output variable that, for each year, holds the distribution of cash flow

NPV is an output variable that holds the distribution of project net present value

NPVSIMPLE is a system variable that calculates NPV

tax_rate is a variable that holds the corporate tax-rate coefficient

ror is an input variable that holds the distribution for required rate of return

Other variables used but not in the equations in this part include the following.

play_risk is a chance of failure variable that holds the chance that the project will fail due to misinterpretation of information

tech_risk is a chance of failure variable that holds the chance that the project will fail due to lack of appropriate technology

market_risk is a chance of failure variable that holds the chance that the project will fail due to market forces

contract_risk is a chance of failure variable that holds the chance that the project will fail, before inception, due to contractual disputes

political_risk is a chance of failure variable that holds the chance that the project will fail, before inception, due to political problems

envir_risk is a chance of failure variable that holds the chance that the project will fail, before inception, due to environmental problems

POPULATING THE MODEL WITH DATA

Geophysicists and geologists have indicated to Ivan that the average field size is relatively small. Their estimates range from 0.5 million barrels to 5 million barrels per field in the play trend. They are relatively confident that the average field size will be about 2 million barrels. To reflect their information and intuition, Ivan and Glenn assign a peakedness of 8 to the distribution. The resulting frequency plot is shown in Figure 5.2.

Geologists and geophysicists also have taken a best guess at the number of fields that exist in the trend. Because these fields are structural in nature and, therefore, more easily defined, the scientists believe that about 10 fields of sufficient size to be economic exist in the play trend. The range of 8 to 13 reflects their uncertainty regarding the size and, thereby, the economic viability of some of the subsurface structures. Because of the ease of identification of the structures, a peakedness of 10 was assigned to the distribution of the number of fields. The resulting plot is shown in Figure 5.3.

The number (percent) of fields that can be drilled per year is basically dictated by funds, equipment availability, and weather. Considering all factors, Ivan believes that they can drill 50% of the total number of wells in the first year, 20% of the wells in the second year, and 10% of the wells in each of the remaining 3 years.

Due to the viscous nature of the oil and the permeability of the reservoir rocks, recovery factors for the area are known to be relatively low. Geologists believe that, at best, 50% of the oil in the reservoir can be brought to the surface. Worst-case and most likely estimates are 20% and 30% respectively. Because they have drilled wells in this area previously, they are confident in their most likely value, thus, the peakedness of 10. The plot of recovery factor is shown in Figure 5.4.

Prices for oil and oil products are expected to be depressed for at least the next two years or so. Always optimistic analysts project a slight rise in prices after the

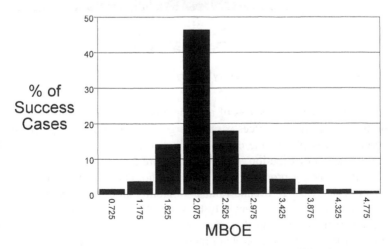

FIGURE 5.2 Frequency plot of average economic field size for oil in place.

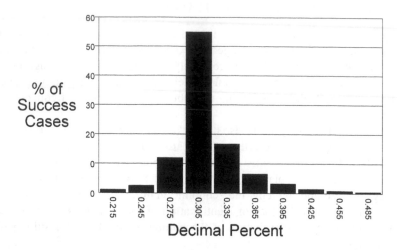

FIGURE 5.3 Frequency plot of the number of economic fields success trend potential.

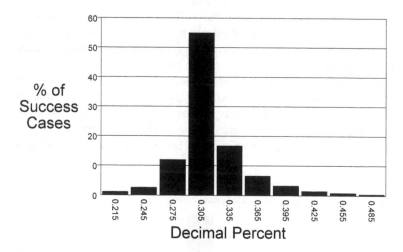

FIGURE 5.4 Frequency plot of recovery factor.

initial depression. The time-series of analyst-supplied prices is shown in Table 5.2. Variability in the projected prices is quite large — from 20% too optimistic to 50% too conservative. A plot of the expanded time-series of prices is shown in Figure 5.5. The expansion distribution is shown in Figure 5.6.

Sales of secondary products such as sulfur are expected to generate relatively small, but ever-increasing, revenue values. Variability around the projected incremental income yearly values is expected to be somewhat less than that projected for oil prices. The incremental income data are shown in Tables 5.2 and 5.3, and the expanded time-series of incremental-value-chain income is depicted in Figure 5.7. The current required rate of return for the corporation is 13%. Corporate tax rate is 50%.

TABLE 5.2
Risk Model Time-Series Data

Variable Name	Value				
	Year 1	Year 2	Year 3	Year 4	Year 5
% Fields Discovered per year (Decimal %)	0.5	0.2	0.1	0.1	1
Current Oil Price ($ per barrel)	12	10	12	13	15
Income from Secondary Product (Thousands $)	100	120	140	170	190
Equipment/Infrastructure Costs (Millions $)	10	2	0.5	0.5	0.5
Geologic/Geophysical Costs (Millions $)	5	1	1	1	1
Personnel Costs (Millions $)	10	5	5	5	5

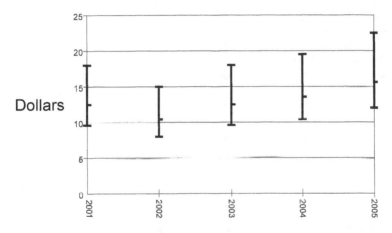

FIGURE 5.5 Time series of expanded prices per barrel.

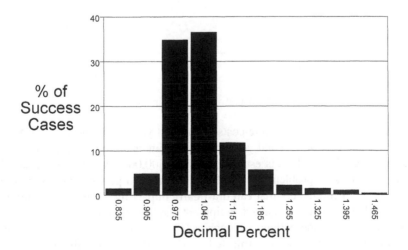

FIGURE 5.6 Expansion distribution for prices per barrell.

TABLE 5.3
Risk-Model Input Distributions

Variable Name	Minimum Value	Most Likely Value	Maximum Value	Peakedness Value
Average Economic Field Size (Millions of barrels)	0.5	2	5	8
Number of Successful Fields	8	10	13	10
Recovery Factor (Decimal %)	0.2	0.3	0.5	10
Variability in $/Barrel (Decimal %)	0.08	1	1.5	10
Variability in Secondary Value Chain (Decimal %)	0.8	1	1.3	10
Required Rate of Return (%)		13		
Corporate Tax Rate (Decimal %)		0.5		
Variability in Equipment/ Infrastructure Costs (Decimal %)	0.5	1	1.7	10
Variability in Geological/ Geophysical Costs (Decimal %)	0.8	1	1.5	10
Variability in Personnel Costs (Decimal %)	0.09	1	1.1	10

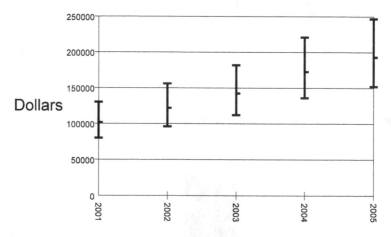

FIGURE 5.7 Time series of incremental value chain income.

Equipment and infrastructure costs in the first year of the project are projected to be about $10 million. Second-year costs are estimated to be about $2 million with remaining years' annual costs estimated to be $500,000. Weather (good and bad), a spate of or lack of problems with subcontractors, and performance of equipment are but a few of the factors that can drastically impact these costs. Ivan and Glenn determine that the yearly projected equipment and infrastructure costs could be high by as much as 50% and perhaps as much as 70% too conservative. This estimate of variability, when applied to the projected yearly costs, results in the time-series plot shown in Figure 5.8.

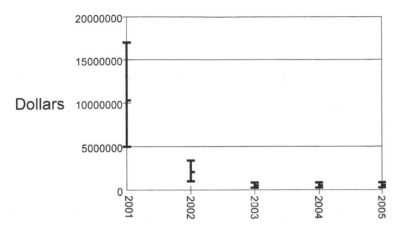

FIGURE 5.8 Time series of expanded equipment and infrastructure costs.

Costs for seismic surveys, geological reconnaissance, and geochemical studies are fairly well known for such projects. Ivan anticipates that costs associated with these efforts will be about $5 million in the first year and about $1 million in each of the remaining 4 years. Historic precedence has shown that these estimates may be exaggerated by as much as 20% or too conservative by as much as 50%. The time-series of expanded costs for geological/geophysical/geochemical work appears in Figure 5.9.

Like the costs for geotechnical work, the expenses associated with payroll for such projects are relatively transparent. Because this model is meant to give only a high-level assessment of the project, personnel costs associated with construction (CAPEX) and with operation of the fields (OPEX) will not be differentiated. Glenn and Ivan know from experience that combined personnel costs for the project are

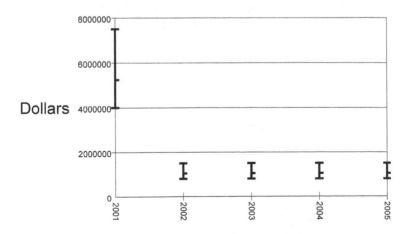

FIGURE 5.9 Time series of expanded geologic/geophysical costs.

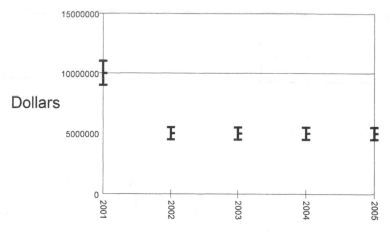

FIGURE 5.10 Time series of expanded personal costs.

likely to be about $10 million in the first year and about half that amount in the remaining years. They believe that these estimates will be accurate to within plus or minus 10%. The expanded time-series of annual personnel costs is shown in Figure 5.10.

Ivan and Glenn have identified six chances for abject project failure. These are shown in Table 5.4. Although the seismic data are of high quality and the structural traps are relatively easily defined, there is a possibility that the rock type intended to serve as reservoir rock contains too high a concentration of shale to be an effective reservoir. Based on anecdotal evidence from other wells and fields in the area, Ivan and Glenn believe that there exists a 20% chance that the reservoir rock of the fields will be of too poor a quality to produce economic volumes of oil.

In complex plays, there is always some chance that reaching and producing the oil will require some technology that is not readily or economically available. Because the traps in the fields being considered present no complications, Ivan and Glenn consider the chance of technical failure to be zero.

TABLE 5.4
Chance-Of-Failure Variables

Variable Name	Chance of Failure (%)
Play Risk	20
Technical Risk	0
Market Risk	0
Contract Risk	10
Political Risk	5
Environmental Risk	10

Risk of failure from market forces can sometimes be a credible threat. In some situations in which sale price, for example, determines whether a local market will consume the discovered resources, too high a price can cause exclusion of the product from the local market. In the case considered here, however, the resources will be sold to the worldwide spot market and, therefore, the risk from market failure is considered to be nil.

Failure from contracts is a different matter. Some of the land in the trend of fields is owned by groups with whom, in the past, favorable contract terms could not be reached. Ivan believes that he has good rapport with these groups and that a reasonable contract agreement can be reached. However, he will allow a 10% chance of failure associated with contract problems.

Political problems also threaten the project. Government bodies have in the past refused to issue required permits to companies with whom the government has run afoul. B&R has not had direct problems with the government, but B&R has been part of joint ventures with other companies known to be on the government's "hit list." The threat from the government is small, but real. Ivan assigns a 5% chance of failure to this parameter.

Finally, the threat from environmental groups must be considered. B&R has a stellar environmental track record, but engineers have indicated that in the field-trend area, disposal of liquid waste from the drilling process might be a problem. Based upon the estimate from the engineers, Ivan and Glenn agree that there is about a 10% chance that environmental groups might mount a campaign vigorous enough to halt the project.

RESULTS FROM MODEL EXECUTION

As shown by the cumulative frequency plot of total resources in Figure 5.11, reserve levels range from about 1 million to around 20 million barrels. There is a 20% chance of failure due to poor reservoir rock, as evidenced by the cumulative frequency curve's Y-axis intercept value of 80. The economic limit for this plot, expressed in barrels, is set at 5 million. This number was generated by a "back of the envelope" calculation based on the Total Cost curve shown in Figure 5.12 and an average oil price of $14 per barrel.

The Total Cost curve in Figure 5.12 indicates no chance of failure (i.e., cumulative frequency curve goes to 100 on the Y axis). No chance of failure is considered for costs because, Ivan and Glenn have reasoned, at least some first-year costs might be realized before a particular failure mode would indicate abject failure of the project. Figure 5.12 also shows that the net risk-weighted cost for the project is about $7.2 million. With an average oil price for the project of $14 per barrel, the duo determined that they would have to recover about 5 million barrels just to break even (not considering a required rate of return), thus, the 5-million-barrel economic limit.

From the Total Resources plot in Figure 5.11, it can be seen that the economic risk-weighted resources value is about 5.4 million barrels. This figure reflects the 20% chance of total failure and the 5-million-barrel economic limit. Because the

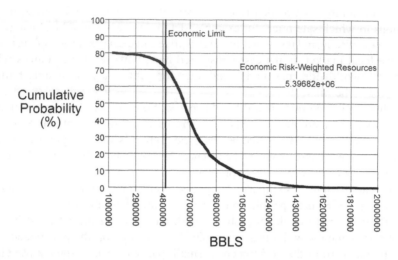

FIGURE 5.11 Cumulative frequency plot of total undiscovered resources.

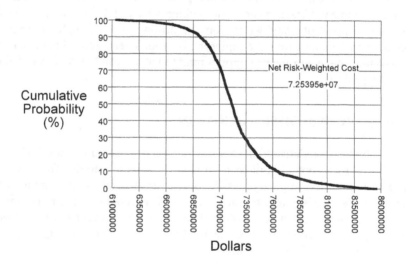

FIGURE 5.12 Cumulative frequency plot of total cost for project.

risk-weighted resources for the project are just slightly greater than the economic limit, Ivan and Glenn are becoming a bit pessimistic about the project.

From the Undiscovered Resources per Year time-series plot in Figure 5.13, Ivan and Glenn note that the resources, as expected, decline with time. They also note from this plot that the minimum value (bottom end of a vertical bar) for each year approaches zero.

A time-series plot of Revenue (Figure 5.14) mimics that of Undiscovered Resources per Year in that the minimum revenue each year comes close to zero. Unlike Resources per Year, however, in later years the mean (middle horizontal bar on the vertical bar) and the maximum values for each year (top of each vertical

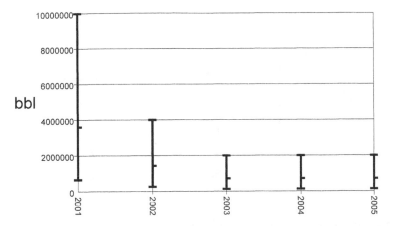

FIGURE 5.13   Time series plot of undiscovered resources per year.

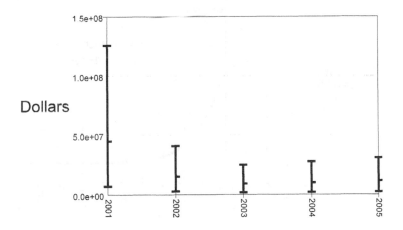

FIGURE 5.14   Time series plot of revenue.

bar) rise slightly with time. This is explained by the projected rise in oil prices through time.

As shown in Figure 5.15, yearly cash flow has a significant chance of being negative. In later years, the mean cash flow value is just above zero. A frequency plot of third-year cash flow (Figure 5.16) shows the proportion of negative and nonfailure values.

As expected from inspection of the cash-flow plot, the NPV cumulative frequency curve has a significant negative portion (see Figure 5.17). In addition, the chance of failure has increased due to inclusion of chance of failure from contract, political, and environmental factors. The following is the new Total Chance of Success (TCS).

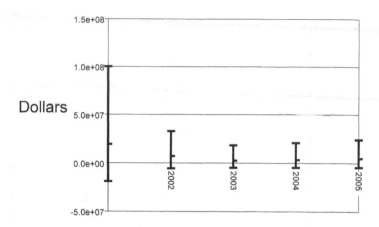

FIGURE 5.15 Time series plot of yearly cash flow.

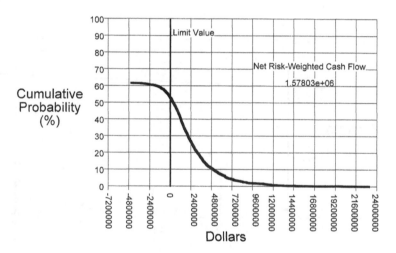

FIGURE 5.16 Time series plot of third-year cash flow.

TCS = (1 − Chance of failure from play risk) × (1 − Chance of failure from contract risk) × (1 − Chance of failure from political risk) × (1 − Chance of failure from environmental risk) =

(1 − 0.2) × (1 − 0.1) × (1 − 0.05) × (1 − 0.1) =

0.8 × 0.9 × 0.95 × 0.9 =

0.6156 × 100 =

61.56%

Considering the economic limit and the TCS, there exists only about a 58% chance that this project will result in a positive NPV. The Net Risk-Weighted value is only about $10 million. Although this risk-weighted value is positive, it is small relative to the costs of the project. Ivan and Glenn determine that this project likely

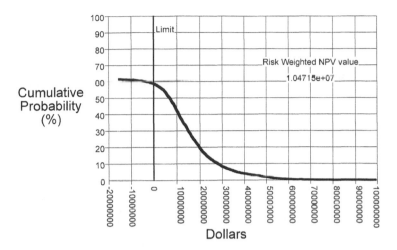

FIGURE 5.17 Cumulative frequency plot of net present value of opportunity.

will only be pursued if other, more profitable projects do not present themselves. They are, however, pleased with the ease of execution and output from the new risk model and will use it on all subsequent evaluations.

SO, SO WHAT?

In previous and subsequent examples in this book, reasons for considering use of a risk/uncertainty model are primarily financial in nature or stem from unsatisfactory results from a previously employed process. This example puts forth the precept that switching to the use of such modeling techniques might be driven by changes in scope or volume of business and not necessarily because of any dissatisfaction with the results from the old process. In this example, Ivan was not dissatisfied with either the process or the results of his mode of operation; however, increases in requests for funding dictated that something "had to give." Thus, the consideration of the consistent-modeling approach.

Another point to be made is that switching to such a new process can significantly change the way business is done. For example, Ivan can no longer afford the luxury of visiting each potential client. This is a significant change that must be dealt with, both physically and psychologically. In addition, Ivan must now develop a feeling of trust that the new way of doing things sufficiently accounts for the feelings and judgments that he made during his visits. It is typical when switching to a quantitative technique to have to find a way to incorporate "intangible" concepts such as Ivan's impressions resulting from a visit.

It also is important to focus on value and to consider chances of failure that will have a significant impact on the value of projects. "Soft" issues (see Chapter 6) such as impressions and judgments can often be incorporated in an analysis as chances of failure and, therefore, can be used to impact the result.

6 Using Chance of Failure and Risk-Weighted Values to Reflect the Effect of "Soft" Issues on the Value of an Opportunity

CONTENTS

INTRODUCTION

Regardless of whether you take a scat-of-the-pants or high-tech tack, each opportunity (in your portfolio) must be assigned a value or range of values. Whether these values are perceived or calculated or divined by mystic powers, before you can compare/rank/select (i.e., portfolio manage) a collection of opportunities, you must know the potential benefit or liability represented by each opportunity. This chapter will cover how to determine the magnitude of that benefit or liability for each item and the collective worth of an entire portfolio of opportunities.

Uncertainty concerning a risk/uncertainty-model parameter can be represented by a distribution of coefficients (although distributions certainly are not the sole means of expressing uncertainty). The distribution of coefficients represents a range of values for the parameter. (See Chapter 15 for a full discourse on distributions.) Those accustomed to feeding data to spreadsheets or those new to the process of thinking about what value or values might best represent a parameter typically find it difficult to contemplate more than one value that might represent a risk/uncertainty-model element (i.e., a range). Sometimes it is difficult for some people to make the transition from a deterministic (single-valued) mindset to one that accommodates arrays of values for a variable.

Moreover, if the transition from thinking about single values to considering ranges is vexing, then the process of coaxing people to realize and express (two different things) the full and real range of coefficient-related values is almost completely unfathomable. It is probably human nature and certainly human tendency to misrepresent just how "good" or "bad" a given situation realistically might be.

For example, it might be the charge of a group to estimate the expense associated with the repair of a machine's component while the machine is taken off-line for scheduled maintenance. The group might be cognizant of the cost of repairing the component in the past. Therefore, it is not uncommon for the group to fixate on a single value or small range of values to represent the cost of component repair.

Because the component is scheduled for repair and because the group has always in the past refurbished the component, it might be a nearly impossible undertaking to coax the group into considering that the component might not be in need of refurbishment and that the cost could be zero. Conversely, in the past the component has been only repaired, not replaced. A worthy discussion facilitator should get group members to consider that the part might require replacement. If a seriously damaged part has adversely affected adjacent parts, those parts might also require repair or replacement. However unlikely, therefore, costs for component repair could range from zero to several times the traditional cost. As previously stated, it is a daunting task to prompt people first to realize and then to verbally express (admit) the full range of possibilities.

If the facilitator has done a credible job with regard to building truly representative distributions for the "measurable" parameters, a risk/uncertainty model can be constructed that is capable of calculating the value of an opportunity (or liability) at certainty. That is, the cumulative frequency curve representing the range of model-output values on the cumulative frequency plot will range from a probability near zero to one at 100%. This means that there is a 100% chance of achieving the minimum X-axis value or a value of greater magnitude (see Fundamentals of Risk Assessment section of this book).

There exist, however, concerns and issues, the effects of which are not best captured as distributions. These items are referred to as "soft" issues. It is absolutely critical to account for the impact of soft issues on the value of an opportunity. Consideration of soft issues is critical because we need to know not only what the opportunity would produce if it is successfully executed, but also what a given opportunity would contribute to a portfolio of opportunities on a risk-weighted basis. This, as delineated in the remainder of this chapter, is accomplished by using soft issues to impact the Y-axis (probability) intercept of the cumulative frequency curve and by calculating one of a host of risk-weighted values.

Although soft issues can express themselves in a multitude of ways, we will here limit our consideration to the type of factor that could cause an element of a portfolio to be dropped from consideration for execution. The following are examples of such soft issues (i.e., chances of failure).

- Environmental — An environmental group might raise objections significant enough to cause a proposed project to be dropped from the portfolio of potential projects.

- Political — Political support for a project might be withdrawn, or political considerations between countries might change in a manner that causes a project to be considered untenable.
- Financial — Financial backing for a project could be withdrawn, causing the project to be dropped from the portfolio of possible projects.
- Management — Management backing for a project could wane or be completely withdrawn.
- Partners — Support for a project from a partner could fade, requiring that the project be dropped from the portfolio.
- Legislation — Zoning laws could change and preclude a project.

ACCURATE ESTIMATES OF VALUE ARE ESSENTIAL

In business, time is money. If a commodity cannot be monetized in a timely fashion, often the window of profit-making opportunity is lost. Therefore, it is as big a mistake to underestimate volume or value of a commodity as it is to overestimate it.

For example, the manager of a factory that produces perishable goods might project that next month the factory can manufacture 1000 units. In response to the production projections, the marketing, distribution, transportation, and other departments plan to handle 1000 units. The transportation department, for example, signs contracts with trucking firms and the railroad to ship 1000 units. In the end, however, the factory produced 1200 units in the month. The extra 200 units can only be converted to profits if they have been contracted for by retail outlets and can be transported to the outlets. If transport contracts cannot be augmented, then the extra 200 units will return no value (in fact, will have negative value) because they cannot be brought to market — the result of underestimating production.

TYPES OF CHANCE OF FAILURE

In business, failure can be broadly categorized as two types: financial and abject. We will first consider here an overview of financial failure. In-depth explanations of chance of failure and examples are given in the previous book.

A probabilistic financial analysis can result in both positive and negative values. In the simplest form, negative values result when expenses are greater than income (see Figure 6.1). Figure 6.1 shows that the analyzed opportunity has about a 15% chance of financial failure (values less than zero, although in a real analysis the financial-success threshold can be significantly on either side of zero). That is, we expect there is a 15% chance that this opportunity will produce a return we consider to represent financial failure of the project. Whether a project returns a positive or negative value, financial failure (or not) implies that the project was actually selected and executed.

The concept of abject failure is different. When implementing abject failure, we typically (not always) are attempting to assess situations that could cause failure of the project before inception of execution. For example, if we plan a car trip during which we are to evaluate the car's fuel efficiency but find that the car has a flat tire, the trip is canceled and the evaluations not made. Similarly, we might plan to build

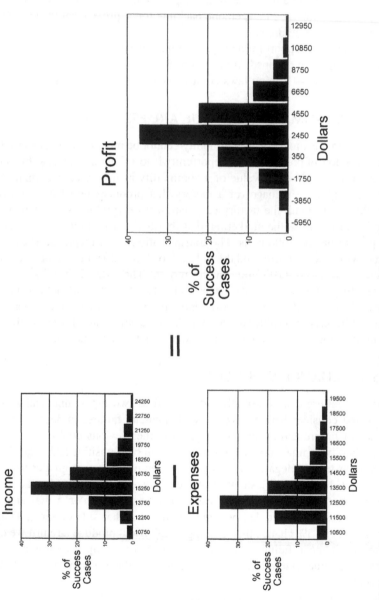

FIGURE 6.1 Chance of financial failure: Simple Monte Carlo analysis plot of income, expenses, and profit (calculated by subtracting expenses from income) showing possibility of both positive and negative profit.

a chemical plant. Before we can begin, however, the land on which the plant was to be built is zoned residential and the project canceled.

Corporate executives generally concern themselves with only the chance of financial failure (about 20% in this case). This is typical. Incorporation of abject failure changes the decision-making process entirely. As shown in Figure 6.2, abject failure modes are like links in a chain. It takes only one failure to "kill" a project.

Let us consider that the simple plots in Figure 6.1 represent our estimates of income and expenses associated with construction and operation of a new manufacturing facility. We might discover that there are three factors that could kill the project before it gets started (i.e., three chances of abject failure).

First, the competition might announce and/or begin construction of a similar facility. The market cannot absorb the production from two such facilities. We believe there is a 15% chance that the competition will begin construction before we do and, therefore, we will not pursue the project.

Second, the property on which we plan to build our facility might be zoned as residential rather than commercial. If this happens, we cannot build our facility and the project will simply be dropped from our portfolio of opportunities; the project will contribute nothing to the portfolio value. There is a 20% chance of zoning problems.

Finally, objections from environmental groups could block construction. We estimate a 10% chance of environmental problems.

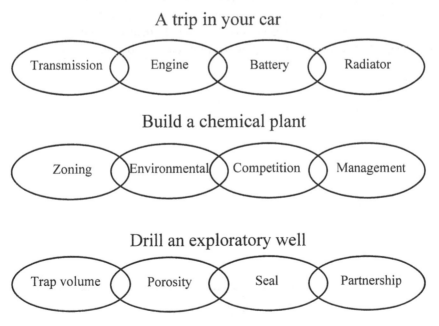

FIGURE 6.2 Plot showing that abject failure modes are similar to links in a chain — only one has to fail for the project to fail.

Total chance of success should be calculated as follows.

Total chance of success (TCOS)
Chance of failure of variable N (COF N)
TCOS = (1 − COF competition) * (1 − Zoning) * (1 − Environmental) =
    (1 − 0.15) * (1 − 0.2) * (1 − 0.1) = 0.612
    0.612 * 100 = 61.2% Total chance of success

HOW TO EXPRESS AND USE CHANCE OF FAILURE

So, what do we do with this total chance of success? Again, details of how chance of failure, risk-weighted values, and cumulative frequency curves work together are given in the precursor to this book. Only a small portion of the complete explanation will be given here.

In Figure 6.3, eight dots are arranged along the X axis of the plot. For the sake of this example, we will assume that these eight points represent the results of eight calculations of profits from eight consecutive Monte Carlo iterations.

To create a cumulative frequency curve, we divide the Y axis into eight equal segments (number of segments equal to the number of iterations). If all eight dots were labeled with their X-axis values and put into the proverbial hat, the probability of drawing the dot with the X-axis value of greatest magnitude is one in eight. Therefore, on a cumulative frequency plot, the right-most point is raised 1/8 of the way up the Y axis. The probability of selecting from the hat the dot representing the greatest magnitude X-axis value OR the dot representing the second greatest X-axis value is 2 out of 8. The dot second from the right end is therefore raised 1/4 of the way up the Y axis. All subsequent dots are similarly treated, resulting in the cumulative frequency curve shown in Figure 6.4 (dots connected with lines in Figure 6.5).

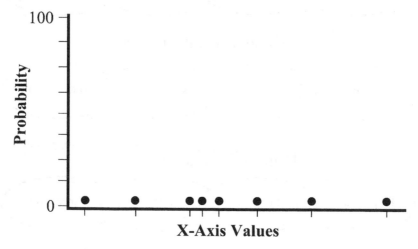

FIGURE 6.3 Chance of abject failure plot of 8 values resulting from 8 Monte Carlo iterations of the production facility.

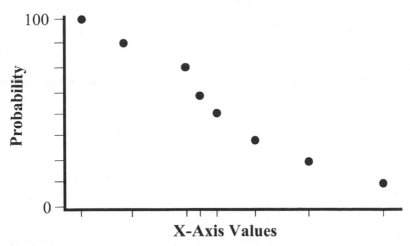

FIGURE 6.4 Chance of abject failure plot of 8 values resulting from Monte Carlo process raised one probability unit to their appropriate probability positions resulting in the cumulative frequency curve.

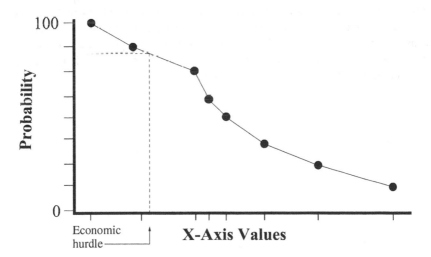

FIGURE 6.5 Chance of abject failure plot of 8 values resulting from Monte Carlo analysis of the production facility raised to their appropriate probability positions and connected to form a cumulative frequency curve.

Let us assume, for the sake of this example, that the X-axis values on our plot represent the number of dollars of profit that might be realized by a company. The "economic hurdle" indicated on the X axis is the product-production limit that must be realized by the company for the project to be deemed successful. Examination of the plot indicates that there is about an 80% chance that the company will realize minimum production represented by the indicated hurdle rate (i.e., about an 80%

chance of realizing the minimum production limit or a value greater than the hurdle value). Conversely, there is about a 20% chance of financial failure. We calculated that our total chance of success for the project was about 61%. What we need to do now is to adjust the plot in Figure 6.5 to account for and reflect the probability of project success.

To account for the approximate 39% chance of abject failure, we depress the Y-axis intercept of the cumulative frequency plot as shown in Figure 6.6. In Figure 6.6, the Y-axis intercept is about 61% of the way "up from the bottom" of the Y axis and about 39% of the way "down from the top" of the Y axis (percents of success and failure, respectively). Now, we extend a line from the minimum production value on the X axis "up to" the cumulative frequency curve and then project that intercept point to the Y axis. It can be seen from Figure 6.6 that now, having considered abject failure, our chance of achieving a minimum acceptable production value is about 50%. That is, the combined chance of abject and financial failure gives us only a 50% chance of project success.

The question arises as to why we do not simply incorporate the chance of abject failure into the values displayed along the X axis rather than depressing the cumulative frequency curve's Y-axis intercept by this amount. We do not do this for several reasons.

First, when calculating the chance of abject failure, we are attempting to separate from the truly production/financial aspects (X axis) the problems that could cause the project to fail before execution of the project has begun. It is difficult, if not impossible, to estimate the production/monetary effect of pre-execution failure scenarios. The X-axis values represent the benefit of the project *IF* it is selected for execution and actually executed. That is what we want.

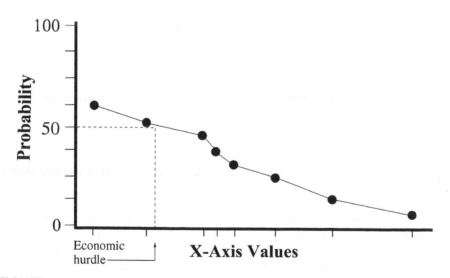

FIGURE 6.6 Cumulative frequency curve's Y-axis intercept position adjusted for the total chance of success at 61%.

In addition, a portfolio can be composed of multiple opportunities that have yet to be executed. The value of that portfolio has been estimated (calculated); however, some of the opportunities in the portfolio simply will not be pursued. That is, they will not get a chance to make or lose money. We do not want to contaminate X-axis financial values with the effects of pre-execution failure (even if we could deduce a means to do that). We would like the X-axis values to represent the amount of benefit the project would produce *IF* it is executed.

Furthermore, in our example of the hypothetical manufacturing facility, our plot of simple profit (Figure 6.1) shows that the project will most likely produce about $2000 to $4000 in profits *IF* the project is selected for execution. Incorporation of the chance of abject failure into the X-axis values would destroy our ability to know how much a project would produce if executed.

A second example might be a probability plot in which the X-axis values represent barrels of oil that a prospect might produce. Our engineers need to know how large a production facility they have to build to handle the volume of oil from the prospect *IF* it produces oil. Reducing the X-axis values by the chance of abject failure might cause our engineers to plan a facility that is incapable of handling the volume of oil produced by the prospect. Similar miscalculations can be made regardless of the X-axis units.

So, let us assume that we should consider the effect of chance of failure. Let us also assume that we should not incorporate chance of abject failure into the X-axis-value calculations but, rather, should depress the cumulative frequency curve's Y-axis intercept by the combined chances of abject failure. How do we actually use this information to determine the value of a portfolio? The answer is, through the use of risk-weighted values.

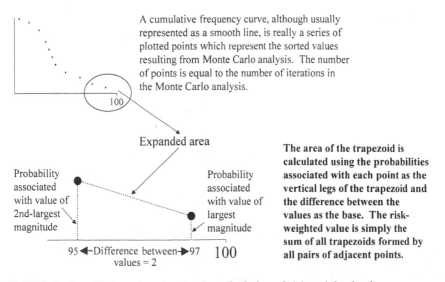

A cumulative frequency curve, although usually represented as a smooth line, is really a series of plotted points which represent the sorted values resulting from Monte Carlo analysis. The number of points is equal to the number of iterations in the Monte Carlo analysis.

Expanded area

Probability associated with value of 2nd-largest magnitude

Probability associated with value of largest magnitude

95 ◄–Difference between–► 97   100
values = 2

The area of the trapezoid is calculated using the probabilities associated with each point as the vertical legs of the trapezoid and the difference between the values as the base. The risk-weighted value is simply the sum of all trapezoids formed by all pairs of adjacent points.

FIGURE 6.7 Simplified explanation for the calculation of risk-weighted values.

RISK-WEIGHTED VALUES AND THE VALUE OF A
PORTFOLIO ELEMENT

A complete treatment of the concept of risk-weighted values is given in the Fundamentals section of this book. Readers are directed to that volume for an in-depth
treatment of the subject. The purpose of risk-weighted values is to integrate into a
single representative value a cumulative frequency curve's X-axis values, curve
shape, X-axis thresholds, chance of failure, and other attributes. A rudimentary
explanation regarding the calculation of risk-weighted values in general is shown in
Figure 6.7.

   As delineated in the Fundamentals section, risk-weighted values can be defined
and calculated in a variety of ways (net risk-weighted value, economic risk-weighted
resource value, above a limit value, etc.). In this example, we will concern ourselves
with only the net risk-weighted value. A cursory explanation of the net value is given
in Figure 6.8. Now let us look at our new-manufacturing-facility example depicted
in Figure 6.9. The cumulative frequency curve in this example is composed of 1500
individual points that were sorted on the X axis. Each point, beginning at the right
end of the plot, was raised 1/1500 of the Y-axis range more than the point immediately to its right, resulting in the curve shown in Figure 6.9. f

   The net risk-weighted value for the plot in Figure 6.9 is about 2770. This is also
the mean for the X-axis values. The 50% probability (Y axis) is equal to the 50th

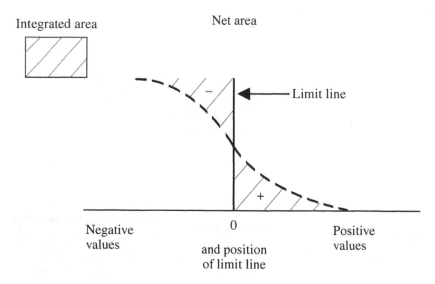

FIGURE 6.8 Simple diagram for calculation of a net risk-weighted value. The limit line in
the Net-area case can be positioned above or below the 0 position. If below 0, the integrated
area includes all the area under the curve to the right of the limit line, even if the line is to
the left of the end of the curve. For a limit line above 0, the area above the curve to left of
the line is itnegrated.

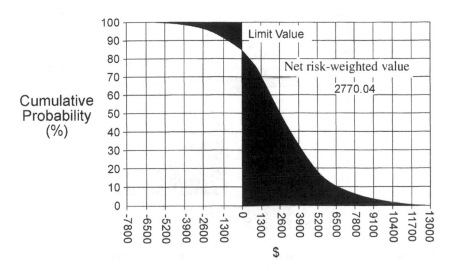

FIGURE 6.9 Simple diagram for calculation of a net risk-weighted value.

percentile. With introduction of the chance of abject failure, none of these things will be true.

Figure 6.10 shows the same data plotted with consideration of the chance of abject failure. The new net risk-weighed value is about 1722. This value integrates financial failure (shaded area to the left of the limit line) and abject failure (Y-axis depression of the cumulative frequency curve). The 1722 value is not the mean of the X-axis values. Probabilities and percentiles are no longer equivalent (what is the percentile for the 70% probability?). In a portfolio of opportunities, this opportunity would be represented by a value of 1722 rather than 2770. In Figure 6.10 the 1722 value, although lesser in magnitude than the 2770 value, still is represented on the X axis. This is not always the case.

Consider the situation of the oil or gas exploration prospect. Such opportunities are fraught with individual chances of failure of geological, geophysical, geochemical, structural, and man-made origin. It is not uncommon for wildcat exploration prospects to sport a total chance of success in the zero to 30% range (i.e., a 70% or greater chance of abject failure). Figure 6.11 is a plot representative of such a prospect.

From Figure 6.11 it can be seen that the total chance of success of this prospect is only about 19%. X-axis values indicate that *IF* this prospect is successful, it will produce no less than about 310 million barrels of oil and no more than around 1600 million barrels. The risk-weighted value for this prospect (i.e., the integrated area under the cumulative frequency curve) is only 188.5 due to the small magnitude of the total chance of success (19%). The mean X-axis value is about 991. The mean value is a reasonable guess as to how much oil this prospect will produce (and, therefore, how big a facility engineers should build to handle the oil produced) *IF* the prospect produces any oil at all (that is, if it is not a "dry hole"). There is an

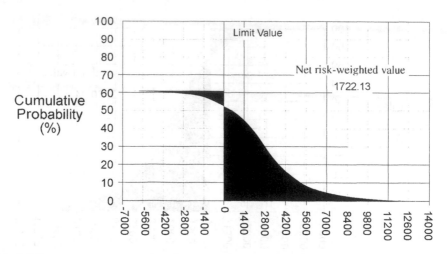

FIGURE 6.10 Cumulative frequency plot from Figure 6.9 showing effect of considering abject chance of success. The new risk-weighted value of about 1722 integrates financial (shaded area to the left of the limit line) and abject (Y-axis depression) failure.

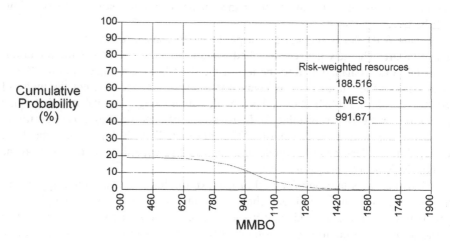

FIGURE 6.11 Cumulative frequency plot for volume of recoverable oil from a prospect showing that the risk-weighted value is outside the range of the cumulative frequency curve. Exploration prospects often are represented in a prospect portfolio by a risk-weighted value that is smaller than any X-axis value that can be produced by combination of the input distributions, but this is all right.

81% chance that the prospect will contribute no resources (no oil or gas) to a portfolio of, say, 100 prospects.

Engineers planning a production facility would use the 991 value to plan their production facility (rather than the 188 value) because *IF* this prospect is successful, the 991 figure represents the amount of oil with which they will have to contend.

However, corporate forecasters attempting to determine how much oil this prospect would contribute to a portfolio of prospects on a fully risk-weighted basis would use the 188 value. Note that the 188 value is not even a possible X-axis value. That is, the prospect is not even capable of producing an amount of oil as small as 188. That is fine.

If we had a hypothetical portfolio of 100 prospects exactly like this one, the value of the portfolio would *not* be

$$100 \times 1020 = 102,000.$$

The value of the portfolio, given the significant chance of abject failure of each opportunity (about 81% chance that each element of the portfolio will contribute nothing to the portfolio value), would be the sum of the 100 risk-weighted values or

$$100 \times 188 = 18,800 \text{ barrels.}$$

Thus, in reality, the 18,800 figure represents the likely value of the portfolio. The practice of simply substituting zeros for failure is a bad one. See a detailed discussion of this in the earlier book.

Let us now return to our manufacturing facility example in which we have about a 39% chance of abject failure (61% chance of success not considering financial failure). In Figure 6.10, the depression of the cumulative frequency curve's Y-axis intercept represents the chance of abject project failure and the shaded area to the left of the zero X-axis value indicates the chance of financial failure. The net risk-weighted value integrates all failure, the curve shape, and other attributes to arrive at a net risk-weighted value of about 1722.

The value of a hypothetical portfolio of 100 identical projects would be the sum of all risk-weighted values or

$$100 \times \$1722 = \$172,200$$

rather than

$$100 \times \$2770 = \$277,000.$$

Of course, in the real world, actual assessments involve time-series analyses, generation of cash flows, and calculation of financial measures such as NPV or IRR. Implementation of financial and abject chances of failure and risk-weighted values helps determine the actual value of this project, and, in turn, the value of the portfolio of which it is an element.

VALUE OF A PORTFOLIO COMPOSED OF DISSIMILAR ELEMENTS

In the preceding examples, we considered, for the sake of simplicity, hypothetical portfolios of 100 identical opportunities. It is common, however, for portfolios to be composed of dissimilar elements.

Most large, multinational, and integrated corporations manipulate portfolios that contain entities representing investment opportunities of different types. For example, let us consider a corporation attempting to calculate the value of a portfolio of two opportunities. We will consider one of our portfolio elements to be an oil-field-development opportunity from Chapter 5. The second opportunity in our portfolio will be a chance to construct a new manufacturing facility for the production of special steel bars discussed in Chapter 4.

It should be pointed out here that the question to be answered in this section relates to the value of the portfolio. We will not consider, for example, which elements of the portfolio should be pursued and which should not. The practice of portfolio management is a more complex issue and will not be addressed (see Selected Readings section of this book for references on efficient frontiers and other portfolio-management techniques).

Though somewhat less subjective than portfolio management, determining the value of a portfolio is fundamental to good business. Production and financial projections form the basis for setting stock prices, salaries, and for scheduling manpower and other physical resources. As delineated in previous sections of this chapter, when determining the value of an opportunity, chances of failure — both financial and abject — need to be considered.

The net present value of our oil-field-development project is about $10 million. This value is the result of integration of annual cash flow values and individual chances of failure. The NPV plot in Figure 6.12 shows that the project has about a 61.5% chance of success and about a 58% chance of yielding a positive NPV value. Without considering the chance of failure, the apparent value of this portfolio element would be well in excess of the $10 million figure.

Our second project, a new manufacturing facility for producing steel bars, has a net risk-weighted NPV of $3.3 million. This value results from integrating the

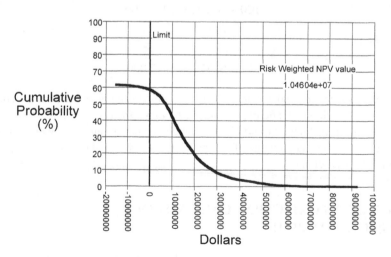

FIGURE 6.12 Cumulative frequency plot of net present value from the oil-field-development opportunity discussed in Chapter 5.

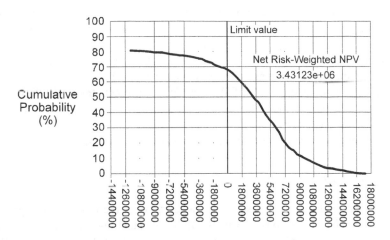

FIGURE 6.13 Cumulative frequency plot of net present value from the price-cycle example discussed in Chapter 3.

annual cash flow values, the 5% chance of economic failure, and the 20% chance of abject failure. The NPV plot for this project is shown in Figure 6.13.

Acknowledging that there exist 38.5% and 20% chances that our oil-field and manufacturing facility opportunities, respectively, will contribute no value to the portfolio, the actual value of the portfolio is the sum of the risk-weighted values as follows.

Total value of the portfolio = (risk-weighted NPV for oil field) +
(risk-weighted NPV for manufacturing facility) =
($10 million + $3.3 million) =
$13.3 million

This value is, of course, based only upon the risk-weighted NPVs of the projects. It should be noted that any risk-weighted financial measure (Internal Rate of Return, Discounted Rate of Return, Return on Capital Employed, etc.) or any combination of risk-weighted values could be used to establish the value of a portfolio.

In this example, as in other examples, we are treating abject failure as though it is cost free. Sometimes this is true. For example, suppose someone on the corporate staff, as part of his or her normal charge, proposes a construction project. That project might then be added to the corporate portfolio of projects that might be pursued. The project has as an integral component an abject chance of failure related to governmental edicts that might be passed that would preclude the project. The NPV added to the portfolio value for the project reflects this and other abject chances of failure.

Suppose further that later that same week, before any action could be taken on the project, the feared governmental restrictions come to pass and the project is no longer viable. The cost of failure, in this case (above and beyond the cost of simply doing business in the normal way) is nil. This, however, is not always the case.

As you might guess, some projects are proposed, valued, added to the portfolio, and acted upon prior to abject failure. The cost of the prefailure work on the project can be viewed in two ways. Some corporations view these costs as sunk costs and thus do not enter them into the calculation of the present value for the project. However, if these costs are not ignored (written off, effectively), then the cost of failure (or the estimated costs for failure at various project stages) might be considered when calculating the value of the project. This is mainly a philosophical decision, but it needs to be made clear that failure is rarely free and that the cost of failure certainly can be accounted for in the value of a portfolio element.

SO, SO WHAT?

Most conventional risk-related software packages do a credible job of representing parameters with distributions of values (although most such packages require the user to supply statistical input such as mean or standard deviation to get the job done — a liability that algorithms I designed have remedied). Distributions created are used to calculate the values that populate the X axis of a frequency or cumulative frequency plot.

It also is the wont of such packages to attempt to incorporate failure aspects, as defined in this book, in the calculation of the X-axis values or to substitute some X-axis value for failure — typically zero is used. This is not prudent practice, because we would like the X-axis coefficients to represent the value of the opportunity *IF* it is successfully executed. As pointed out previously, damping the X-axis values to reflect failure scenarios leads to misleading results regarding the true value of the opportunity.

Incorporation of chance of failure and risk-weighted values as prescribed here allows financial or other results (X-axis coefficients) to be calculated completely independent of failure considerations. Depression of the Y-axis intercept of the cumulative frequency curve and calculation of one of a host of risk-weighted values facilitates the impact of output values without bastardization of X-axis numbers. Such risk-weighted values can be used to represent the contribution of the opportunity to a portfolio of opportunities or can be used as a means of comparing or ranking opportunities. It should be noted that the risk-weighted value is *not* the amount you expect the opportunity to yield (unless the opportunity is considered a certainty). The expected amount can be derived by inspection of the X-axis values (mean, P50, or any other value deemed representative). Consideration of chance of failure in the method described in this book can have a significant impact on the value of an opportunity and, therefore, its value to the corporation.

7 Production-Sharing Agreement Risk Model

CONTENTS

INTRODUCTION

Many contemporary business ventures require significant capital investment. Often it is the case that either the magnitude of the investment required exceeds the amount of money stored in a single corporate coffer or the risk represented by the venture is greater than a single corporation is willing to accept; sometimes it is both. In either or both cases, it is typical for business entities to seek one or more partners with whom the expense and risk can be shared.

When partnerships are established, agreements are created that delineate the responsibilities of and rewards for each party. Documentation of responsibilities and rewards generally takes shape as a Production Sharing Agreement (PSA). In this example, only the reward-related aspects of a PSA in the energy business will be addressed. However, the principles, techniques, processes, and models presented could be applicable to a PSA in any business.

WHAT IS THE QUESTION?

Hershel serves as the business manager at Drainemdry Energy, Inc. He has been instructed to work out the details of a PSA with one of Drainemdry's competitors, Drillemup Energy Co. Hershel realizes that his first task is to build a conceptual framework for the PSA. In preliminary meetings with Drillemup representatives, Hershel has come to understand that some fundamental concerns will have to be addressed by the PSA language and that a pre-PSA-signing risk assessment will have to be performed on the joint venture. The language and risk model will need to consider the following.

1. The development of a robust risk assessment methodology for the purpose of understanding the risks to which the companies will be exposed
2. Modeling the scenarios that will have a negative impact on the return from the PSA, including

   - Unsuccessful or semi-successful exploration results
   - A drop in oil prices
   - Cost overruns
   - Increased payments to governments
   - Changes in government regulations

3. Predicting the venture's upside potential
4. Quantitative analysis of risk and return — whether the PSA matches the companies' requirements for a stable, profitable project

Having done this sort of thing before, Hershel realizes that the risk model to be built will need to be focused, at least at first, on a central issue, the resolution of which considers all of the points in the above list. To facilitate delineation of the fundamental issue to be addressed, Hershel arranges a brainstorming session with representatives from the potential partner company. As always, the point of the brainstorming session is to generate a list of all potential issues. During the brainstorming session, Hershel records the following major questions:

1. What is the volume of recoverable oil?
2. What is the volume of recoverable oil equivalent?
3. What is the volume of oil in place?
4. What are the full-cycle (NPV) economics?
5. What is the volume of recoverable oil considering an economic limit?
6. What will be the daily production rates?

To resolve which of the 6 issues is most profound, Hershel employs the time-honored voting process. To facilitate the process, he lists all six items on a piece of paper and distributes a copy to each meeting member. Hershel then distributes a second paper that is divided into Focus and Not Considered Now sections. The representatives at the meeting are asked to put the number (1 through 6) for each statement into one of the two categories.

The voting completed, Hershel discovers that items 4 and 5 have received the highest number of Focus votes and the number of votes for each question is equal. Hershel then facilitates a long conversation and debate concerning the two issues. At the conclusion of the discussion, the members are again asked to assign the two statements to the Focus or Not Considered Now categories.

Results from the new round of balloting clearly indicate that statement number 4 — What are the full-cycle (NPV) economics — is the top contender. Hershel knows that the next step in the risk-model-building process is to attempt to build a comprehensive contributing-factor diagram. For this multiphase problem, construction of such a diagram will be somewhat complex.

BUILDING THE CONTRIBUTING-FACTOR DIAGRAMS

Complexity arises from the fact that the evolution of this type of PSA involves the consideration of two major parts of the problem. First, the physical/technical aspects of oil exploration and production need to be addressed. After outlining the physical and technical parameters associated with the exploration/production phase, the construction and general financial aspects of the problem must be addressed. Because both of these facets will involve the integration of many variables, Hershel decides to generate two separate contributing-factor diagrams. This, Hershel realizes, will mean engaging two different cadres of bodies in at least two separate contributing-factor-diagram-building meetings.

Hershel convenes the first dual-company meeting with representatives from management, geology, geophysics, geochemistry, and engineering. Because the resulting model will be the template used to evaluate many more such opportunities in the future, Hershel realizes that the contributing-factor diagram will have to be "high level," that is, not too burdened with detail. Following two days of discussion and debate, the team decides to include the following parameters in the physical/technical part of the model.

- Percent oil
- Porosity
- Oil saturation
- Trap volume
- Formation volume factor
- Attic oil
- Volume of oil depleted
- Net to gross
- Recovery factor
- Initial decline rate
- Initial production rate
- Economic limit (daily production — barrels of oil per day — BOPD)
- Hyperbolic decline exponent
- Migration efficiency
- Charge
- Trap timing
- Seal failure
- Trap definition failure
- Reservoir presence failure

Integration of all these parameters will result in the calculation of a distribution of total and yearly recoverable resources for the well. This output parameter will then serve as an input variable for the follow-up model that considers the construction/financial aspects of the project.

Hershel and the team, as previously mentioned, decide to keep the model relatively simple. It is abundantly clear to all involved that parameters such as Production Rates could themselves be broken down into many subvariables. These subvariables

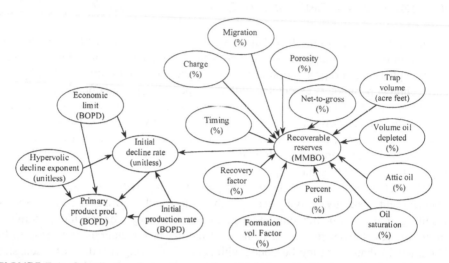

FIGURE 7.1 Contributing-factor diagram for calculation of primary product production.

will not be ignored. Geoscientists and engineers will, outside the risk model, make the appropriate complex calculations and provide to the model only the result of that integration process. This process will be used for nearly all risk-model input parameters. The contributing-factor diagram for this part of the model appears in Figure 7.1.

Hershel now decides to engage the construction engineers and financial analysts in a similar contributing-factor-diagram workshop. The aim of this meeting is to define the variables and intervariable relationships that will result in NPV and internal rate of return (IRR) calculations. After several days of facilitator-lead discussion, the team culminates its meeting with the following list of input variables:

- Recoverable resources (output variable from the first contributing-factor diagram and model)
- Price for oil
- Expansion factor for oil price
- Construction costs
- Expansion factor for construction costs
- Contingency costs
- Expansion factor for contingency costs
- Tangible fraction CAPEX for taxes
- Increased payments
- Expansion factor for 1999 payments
- Expansion factor for 2000 payments
- Expansion factor for 2001-2007 payments
- Expansion factor for 2008-2018 payments
- Shut-down costs
- Government fails to award us the contract
- New controlling party drops the project
- Probability of government regulatory delay

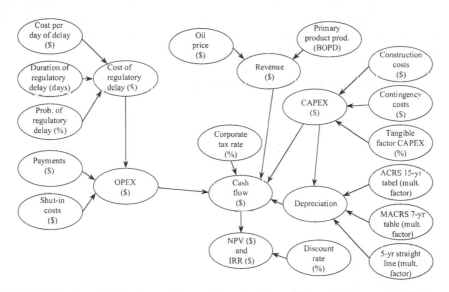

FIGURE 7.2 Contributing-factor diagram for calculation of financial aspects of a PSA.

- Time consumed by regulatory delay
- Cost of regulatory delay
- Corporate tax rate
- Discount rate
- Acclerated Cost Recovery System (ACRS) 15-year table
- Modified Accelerated Cost Recovery System (MACRS) 7-year table
- 5-year straight line

Once again, Hershel and the team realize that the risk model will use as input only high-level variables. Such variables each can be broken down into any number of subcomponents. For example, construction costs could comprise individual costs for pipe, steel, compressors, concrete, exchangers, and the like. The risk model to be constructed will be a time-series algorithm. As such, for certain parameters such as construction costs, users will be required to supply coefficients for each time period (each year, in this case) of the model. It was considered by the team to be unrealistic to expect users to supply yearly values for each of the sub-components of a high-level variable such as construction costs. Therefore, it is expected that appropriate time-dependent calculations will be performed by economists and engineers "outside" the model. The results of those calculations will be input to the model in the form of a time series for high-level variables such as construction costs. The contributing-factor diagram generated by the team appears in Figure 7.2.

RISK-MODEL EQUATIONS

After wrestling with the definition for each variable (see below), the group concentrates on the meaning of the arrows that connect the variables in the two contributing-factor diagrams. That is, the equations that relate the variables need to be agreed upon. Team members write out the relationships among the parameters, and these sentences are supplied to the risk-model-building experts (risk-model programmers) who generate the following risk-model logic.

```
"Convert all inputs from percent to decimal
x=100;
por=por/x;
netgr=netgr/x;
vodep=vodep/x;
attico=attico/x;
oilsat=oilsat/x;
peroil=peroil/x;
rf=rf/x;
spr=spr/x;
migrate=migrate/x;
charge=charge/x;
timing=timing/x;
CapTang=CapTang/x;
"Reserves are entered in MMBOE but need to be bbls for calculations
res1=((7758*por*trapv*netgr*(1-(vodep+attico))*((oilsat*peroil)/fvf)))
    /1000000;
endif
res1 = res1 * migrate * charge * timing;
res = res1 * rf * 1000000;
"Calculate Di as a function of reserves (res), hyperbolic decline exponent (b),
"and economic limit (EL).
Di = ((Qi^b)/((1-b)*res))*((Qi^(1-b))-(EL^(1-b)))*365;
"Calculate yearly production
"When daily production rate drops below EL, stop producing
pp[0]=Qi;
cumpp=0;
cumsp=0;
pp_an = 0;
sp_an = 0;
n=1;
shutin = 0;
while n <_NUM_PER do
"Calculate daily rate
   pp[n]=(Qi*(1+b*Di*n))^(-1/b);
"No longer producing
   if shutin > 0 then
       cumpp[n] = cumpp[n-1];
       cumsp[n] = cumsp[n-1];
       pp_an[n] = 0;
       sp_an[n] = 0;
```

```
    else
"Calculate cumulative prod
        cumpp[n]=((Qi^b)/((1-b)*Di))* (Qi^(1-b) – pp[n]^(1-b))*365;
        cumsp[n]=cumpp[n]*spr;
"Secondary production (gas)
        pp_an[n] = (cumpp[n]-cumpp[n-1])/365;
"Back out annual avg daily rates
        sp_an[n] = (cumsp[n]-cumsp[n-1])/365;
        if pp_an[n] < EL then
            shutin = n;
        endif
    endif
    n = n + 1;
enddo
n = 0;
oilpriceexp = oilprice;
expense = 0;
constcostexp = 0;
contingencyexp = 0;
revenue = 0;
netincome = 0;
atc = 0;
regdelay_total = regdelay_prob * regdelay_time * regdelay cost;
while n <= shutin do
"Note that revenue is calculated from pp_an which is TOTAL production
"(secondary production is neither subtracted from primary nor
"accounted for at a different price).
"$/Bbl
    oilpriceexp[n] = oilprice[n] * shuffle(oilprice_exp);
    revenue[n] = pp_an[n] * 365 * oilpriceexp[n];
"Convert $ to MM$
    revenue[n] = revenue[n]/1000000;
"Op costs
    if n = 0 then
        expense[n] = (payments[n] * shuffle(payments_exp99)) + regdelay_total;
    elseif n = 1 then
        expense[n] = payments[n] * shuffle(payments_exp00);
    elseif n < 8 then
        expense[n] = payments[n] * shuffle(payments_exp0107);
    else
        expense[n] = payments[n] * shuffle(payments_exp0818);
    endif
    if n = shutin then
        expense[n] = expense[n] + ShutInCosts;
    endif
"Cap costs
```

```
    constcostexp[n] = constcost[n] * shuffle(constcost_exp);
    contingencyexp[n] = contingency[n] * shuffle(contingency_exp);
    expenditures[n] = constcostexp[n] + contingencyexp[n];
    n=n+1;
enddo
"Depreciation
"Tangible
Tang7Yr = expenditures * CapTang;
"Intangible
IDCExp = expenditures * (1-CapTang) * .7;
IDC5Yr = expenditures * (1-CapTang) * .3;
"Tangible cannot start until first rate period, but in this model
"first rate period is always period 1
Tang7Yr[1] = Tang7Yr[1] + Tang7Yr[0];
Tang7Yr[0] = 0;
depr = DeprSimple(IDC5Yr,StLine5Tbl) + DeprSimple(Tang7Yr,
    MACRS7Tbl) + IDCExp;
totdepr = 0;
period = 0;
while period <=_NUM_PER-1 do
    totdepr = totdepr + depr[period];
    period = period + 1;
enddo
"Cash flow should also include tariffs, royalties, overhead
noi = revenue – expense;
ti = noi – depr;
if ti > 0 then
    tax = ti * corptax;
else
    tax = 0;
endif
netincome = ti – tax;
cashflow = netincome + depr – expenditures;
"Make sure this was not a dry hole
if cumpp[_NUM_PER-1] > 0 then
"Discounted to mid year
    NPV = NPVsimple(cashflow,discount) / ((discount/100 + 1.) ^ .5);
    IRR = IRRsimple(cashflow,discount);
else
    NPV = -1;
    IRR = -1;
endif
```

where

> **por** is the porosity (pore space) of the reservoir rock (the associated chance
> of failure is that there is no porosity at all)

netgr is the net-to-gross producing/nonproducing reservoir rock ratio.

vodep is the volume of oil already depleted in an already producing reservoir (the associated chance of failure is that all the recoverable oil has already been taken from the reservoir)

attico is the amount of attic oil or "unreachable" oil in the reservoir

oilsat is the percent of the hydrocarbon-filled reservoir that is saturated with oil

peroil is the percent of the reservoir that may be filled with hydrocarbons

rf is the recovery factor which is the percent of the oil in place that can actually be brought to the surface

spr is the secondary product ratio

migrate is the percent of the oil generated by source rocks that actually migrated into the reservoir (trap) rocks (the associated chance of failure is that there was no migration at all — to this trap)

charge is the amount of source-rock-generated hydrocarbons that could reach the reservoir in question (the associated chance of failure is that the source rocks generated no hydrocarbons)

timing is the timing of the formation of the trap relative to the generation and migration of the hydrocarbons (percentages indicate the percent of the hydrocarbons captured due to timing considerations; the associated chance of failure is that the time of trap formation postdated the migration of hydrocarbons)

CapTang is the portion of capital expenditures that is considered tangible for tax purposes.

res1 is the total reserves of the reservoir before application of the recovery factor

trapv is the volume of the trap (reservoir rock)

fvf is the formation volume factor that relates to shrinkage of the trap-rock volume

res is the total reserves after application of recovery factor

Di is the initial decline rate used in the hyperbolic-decline-curve equation

Qi is the initial production rate in barrels of oil per day

b is the value for the "b" term in the hyperbolic-decline-curve equation (normal values of "b" range from 0 to 1)

EL is the economic limit not considering full-cycle economics, the amount of oil that has to be recovered, in barrels of oil per day, for the prospect to be economically attractive

pp is the primary-product production rate at the beginning of each time-series period

cumpp is the cumulative production of the primary product at the end of each time-series period

cumsp is the cumulative production of the secondary product at the end of each time-series period

pp_an is the annual production rate of the primary product for each time-series period

sp_an is the annual production rate of the secondary product for each time-series period

n is a temporary variable used as a period (year) counter

shutin is the period (year) in which the well is shut in (production is halted)

_NUM_PER is a system variable that holds the number of periods in the analysis (number of years over which the analysis will be run)

oilpriceexp is an output time-series of the expanded yearly oil prices

oilprice is a time-series variable for which a deterministic estimate of the price of oil has been supplied by the user for each time period

expense is an output time-series variable that represents for each period the sum of payments and the cost of regulatory delay

constcostexp is an output time-series variable of expanded construction costs

contingencyexp is an output time-series variable of expanded contingency costs

revenue is an output time-series variable that represents for each time period the product of the yearly production and the oil price

netincome is an output time-series variable that represents the difference between revenue and expenses

atc is an output time-series variable that represents the after-tax cash (net income adjusted by corporate tax and expenditures)

regdelay_total is an output distribution of the total cost of a regulatory delay

regdelay_prob is a distribution that represents the percent probability that there will be a government-imposed regulatory delay of the project

regdelay_time is a distribution that represents the duration of a regulatory delay

regdelay_cost is a distribution that represents the cost per unit time of any regulatory delay

shuffle is a system variable that allows, within a single iteration, resampling of a variable for each time period

oilprice_exp is the expansion distribution for oil price (multiplication of each single-valued yearly oil price by this distribution expands the oil price time-series into a series of distributions of yearly oil prices)

payments is a time-series variable for which a deterministic estimate of payments has been supplied by the user for each time period

payments_exp99 is an expansion distribution for payments in the year 1999 (multiplication of the single-valued 1999 payment by this distribution expands the single value into a distribution)

payments_exp00 is an expansion distribution for payments in the year 2000 (multiplication of the single-valued 2000 payment by this distribution expands the single value into a distribution)

payments_exp0107 is an expansion distribution for payments in the years 2001 through 2007 (multiplication of the single-valued 2001-2007 payments by this distribution expands the single value into a distribution)

payments_exp0818 is an expansion distribution for payments in the years 2008 through 2018 (multiplication of the single-valued 2008-2018 payments by this distribution expands the single value into a distribution)

ShutInCosts is the costs associated with shutting in the well

constcost is a time-series variable for which a deterministic estimate of construction costs has been supplied by the user for each time period

constcost_exp is the expansion distribution for construction costs (multiplication of each single-valued yearly construction cost by this distribution expands the construction-cost time-series into a series of distributions of yearly costs)

contingency is a time-series variable for which a deterministic estimate of contingency costs has been supplied by the user for each time period

contingency_exp is the expansion distribution for contingency costs (multiplication of each single-valued yearly contingency cost by this distribution expands the contingency-cost time-series into a series of distributions of yearly contingency costs)

expenditures is an output time-series variable that represents the sum of construction and contingency costs

Tang7Yr is the intangible drilling costs subject to MACRS cost recovery treatment beginning in the year incurred

IDCExp is intangible drilling costs treated as expenses in the year incurred

IDC5Yr is intangible drilling costs subject to 5-year straight-line recovery treatment beginning in the year incurred

depr is a variable that holds the total depreciation value for a given period (year)

DeprSimple is a system variable that calculates depreciation

StLine5Tbl is a table of straight-line recovery coefficients adjusted for mid-year correction with the first time series value to be applied in actual year of spending

MACRS7Tbl is a table of MACRS-recovery coefficients adjusted for mid-year correction with the first time series value to be applied in actual year of spending

totdepr is the sum of all single-period depreciation values

period is a temporary variable used to count time periods

noi is equal to revenue less expenses

ti is taxable income

tax is the amount of tax paid

corptax is a distribution that represents the current corporate tax rate

cashflow is an output time-series variable that represents the sum of net income and depreciation less expenditures

cumpp is a time-series variable of cumulative primary-product production

NPV is an output distribution of net present value

NPVsimple is a system variable that calculates NPV

discount is a distribution that represents the current corporate discount rate

IRR is an output distribution of internal rate of return

IRRsimple is a system variable that calculates IRR

Other chance-of-failure variables that do not appear in the equations because they have no associated distribution include

seal is a chance-of-failure-only variable which represents the chance that the seal on the prospect was (is) insufficient to hold back the rising hydrocarbons

tdf is the trap definition failure — a chance-of-failure-only variable which represents the percent chance that the trap defined does not exist

rpf is the reservoir presence failure — a chance-of-failure-only variable which represents the percent chance that the reservoir described is of the wrong rock type

nocontract is a chance-of-failure-only variable which represents the percent chance that the government will decide not to award the contract to our company

newparty is a chance-of-failure-only variable which represents the chance that a newly elected party will decide to drop the project altogether

POPULATING THE MODEL WITH TECHNICAL DATA

The process of convening two multi-day meetings with two groups representing completely disparate interests (technical and economic) is exhausting enough, but Hershel realizes that he will have to meet with each group again. A second meeting is required to glean from discipline experts the numerical coefficients that will represent each variable in the model.

Prior to such a meeting, however, Hershel must satisfy himself and others that the logic of the risk model which melds the technical and economic aspects of the model has been efficiently and logically composed and that the two parts of the model are succinctly integrated. After receiving the risk-model code from the programming staff, Hershel contacts the principal parties from the technical and economic camps and reviews with them the model logic. Following the all-day meeting, everyone is satisfied that the model captures the essence of the problem. Hershel now feels confident that he can enjoin each group in separate meetings to discuss populating the model with data.

Because the initial part of the model handles the technical facets of the problem, Hershel first convenes a meeting with the technical team. At that meeting, he finds that most of the attendees are not versed in the nuances of probabilistic modeling. Most members of the team have "grown up" utilizing spreadsheets as their primary means of writing and solving systems of equations. They, therefore, have little appreciation or tolerance for generating more than one value per parameter. To build distributions for his probabilistic model, Hershel realizes that he will have to extract from team members, for each parameter, at least minimum, most likely, maximum, and peakedness values. Given the mood and inclination of the attendees, Hershel concludes that it will be impractical to attempt to address every parameter in a probabilistic manner — at least for this first concept-proving run of the model. Therefore, the first order of business in the meeting with the technical cadre is to determine for what subset of the many parameters they will attempt to define distributions. Given the logic of the risk model and the physical nature of the production opportunity, team members agree that only the variables percent oil, porosity, oil saturation, trap volume, formation volume factor, attic oil, volume of oil depleted, net to gross, and recovery factor need to be probabilistically addressed. It is agreed that for each of the remaining parameters that could be represented by distributions (i.e., not chance-of-failure-only variables), single (deterministic) values will suffice for the concept-proving run of the model. This is not what Hershel had hoped for, but he resigns himself to the fact that this response is as good a one as he can reasonably expect. Following a day of debate concerning the coefficients to be used, the team agrees to populate the model with the values shown in Table 7.1.

CHANCES OF ABJECT FAILURE

Hershel was successful in explaining to the technical team the need to include chances of abject failure in the model. These are failure situations that, if they happen, will kill the project outright. For example, the geologists deduced that, given the timing of oil generation and trap formation, a seal rock (over the top of the trap) that had leaked at all over time would have allowed such significant volumes of oil to escape to render the trap uneconomic. Therefore, they deemed the variable Seal quality to be a chance-of-failure-only variable (an on/off switch, in essence). From Table 7.1 it can be seen that they determined a 10% chance that the seal had leaked and, therefore, a 10% chance that the volume of oil would be uneconomic due to seal failure.

Likewise, geologists suggested that porosity should have an associated chance of abject failure. This suggestion was based on the chance that hydothermal subsurface fluids might have destroyed porosity completely (diagenetic effects). For this particular scenario, however, the geologists believe that there is no chance that

TABLE 7.1
Technical Distributions and Chances of Failure

Variable	Minimum	Most Likely	Maximum	Peakedness	Chance of Failure
Percent Oil (Integer %)	30	35	40	5	
Seal Quality (Integer %)					10
Porosity (Integer %)	10	15	20	4	
Oil Saturation (Integer %)	70	80	90	8	
Trap Volume (acre feet)	600,000	800,000	1,000,000	7	
Trap Definition Failure (Integer %)					6
Reservoir Presence Failure (Integer %)					5
Formation Volume Factor	1.1	1.2	1.3	10	
Attic Oil (Integer %)	15	20	25	10	
Volume Oil Depleted (Integer %)	7	10	12	10	0
Net to Gross (Integer %)	65	70	75	10	
Recovery Factor (Integer %)	45	50	55	10	
Initial Production Rate (BOPD)		10,000		10	
Economic Limit (BOPD3		900		10	
Hyperbolic Decline Exponent		0.99		10	
Secondary Production Ratio (Integer %)		30		10	
Migration Efficiency (Integer %)		70		10	0
Charge (Integer %)		100		10	0
Trap Timing (Integer %)		60		10	0

hydrothermal fluids had had an impact on the reservoir — thus the 0% chance of abject porosity failure.

The chance of failure associated with Trap definition failure is the chance that the trap has been misinterpreted from the seismic data and will, in fact, not trap oil. They decided that there exists a 6% chance of abject trap definition failure.

Reservoir presence failure is the chance that the trap is interpreted to be the right structure, but is of the wrong rock type (i.e., the anticline identified actually exists, but is composed of shale not of the hoped-for sandstone). They deemed that this parameter has a 5% chance of abject failure.

For Migration efficiency, the associated chance of abject failure is the chance that either there was no migration at all or that the migrated oil did not migrate toward our reservoir. The chance of abject failure associated with Charge is the chance that the source rocks failed to generate any oil at all. The Trap timing chance of abject failure is the chance that the oil migrated past the position of the reservoir *before* the trap was formed.

Two chances of abject failure were determined for the financial section of the model. Meeting participants agreed that there is some chance that the existing government will fail to award the necessary contract to their company. This, of course, will kill the project outright. Team members agreed that there is about a 15% chance of this happening. In addition, participants determined that there is some chance that a new controlling party will be elected and will drop the project (either before or after initially awarding the contract). They agreed that there is a 10% chance of a newly elected party abandoning the effort.

Given the chance of failure values shown in Table 7.1, the Total Chance of Success (TCOS) is

$$
\begin{aligned}
\text{TCOS} = &(1\text{-seal failure}) \times (1 - \text{trap definition failure}) \times (1 - \text{reservoir presence} \\
&\text{failure}) \times (1 - \text{government contract failure}) \times (1 - \text{new party failure}) = \\
&(1 - 0.1) \times (1 - 0.06) \times (1 - 0.05) \times 1 - 0.15) \times (1 - 0.1) = \\
&0.9 \times 0.94 \times 0.95 \times 0.85 \times 0.9 = \\
&0.6148 \times 100 = 61.48\%
\end{aligned}
$$

This TCOS value will be reflected in the appropriate output plots by depressing the Y-axis intercept of the cumulative frequency curve.

POPULATING THE MODEL WITH FINANCIAL DATA

Hershel next facilitates a meeting with those individuals concerned with the economic aspects of this venture. Because the contributing-factor diagram and program logic have previously been deemed appropriate, Hershel announces at the meeting that the focus will strictly be generating a consensus data set for the model.

Given his experience with the technical group, and realizing that economists also are fond of deterministic spreadsheet-based solutions, Hershel enters this meeting with a realistic set of expectations with regard to the amount and type of data he will be able to gather from the participants. He understands, too, that there is a new twist to the process of gathering the economically related data.

Typically, in deterministic spreadsheets, corporate economists calculate a single cash flow value for each year of 30-year financial projections. This means that for some model input variables, Hershel will have to wrest from the group a deterministic value for each year. This is not so upsetting to meeting participants because they would have to make such yearly extrapolations for their deterministic spreadsheets. In a probabilistic model, however, it is most useful to have each year represented by a distribution. The results of the model are to be net present value (NPV) and internal rate of return (IRR) values. Such parameters are calculated from a series of cash flows.

Hershel knows that extracting the necessary volume of data to generate unique distributions for each year is simply out of the question (for each of 30 values for a single parameter, at the least, a minimum, most likely, and maximum value would have to be supplied — 30 × 3 = 90 input numbers per parameter). As a compromise solution, Hershel proposes that for each time-series variable they will establish just one expansion-factor variable (except in special cases). The coefficients for the expansion factor will, typically, range around 1. The expansion-factor coefficients will be used to generate an expansion factor distribution. On each iteration of the Monte Carlo model, a value from the expansion factor distribution will be selected and used as a multiplier for the annual deterministic value. So, for the Oil price parameter, participants may conclude that their yearly projections for prices may be high by as much as 10% or conservative by as much as 20%. For this particular expansion distribution, a minimum value of 0.9, a most likely value of 1.0 and a maximum value of 1.2 would be used to generate the oil-price expansion factor distribution. So, when a first-year deterministic oil price of, say, $13 is multiplied, on each of the 1500 Monte Carlo iterations, by a value drawn from the expansion-factor distribution, the oil price for that year will be transformed into a distribution of oil prices ranging from $11.70 (10% less than $13) to $15.60 (20% higher than $13). A second-year oil price of, say, $11 would similarly be expanded into a distribution ranging from $9.90 (10% less than $11) to $13.20 (20% more than $11) and so on for each deterministic value for each year.

Hershel has to console himself with this less-than-perfect solution. He knows that applying the same expansion distribution to, for example, oil price values in every year is not optimal. Estimates generally become less certain with increasing time from the present. That is, we may believe that we can predict an average oil price for this year by plus or minus 10% and 20%, respectively. Can we be so sure about 5 years from now? Probably not. Our ability to predict prices 5 years hence likely is no better than plus or minus 50% (or some other relatively large percentage). Therefore, in a more intricate model, Hershel would use a different expansion factor for each year or, at least, a single expansion factor for no more that a few years. For now, Hershel will have to be happy with what he can get.

At the end of the meeting, all members of the group agree on which variables would be handled as time series parameters. These are Oil price, Construction costs, Contingency costs, Increased payments, and three depreciation variables. The values agreed upon for the time series parameters are shown in Table 7.2.

Only the Increased payments variable is selected to have more than one expansion factor. This variable is singled out because the contract with the government,

TABLE 7.2
Time-Series Variables

Years	Oil Price ($/BBL)	Construction Costs 1$ Million)	Contingency Costs ($ Million)	Increased Payments ($ Million)	ACRS 15 Year Table (%)	MACRS 7 Year Table (%)	5-Year Straight Line (%)
1999	13	7	2	2	0.05	0.1428571	0.1
2000	11	4	1	1	0.095	0.244898	0.2
2001	11	3	0.05	0.5	0.0855	0.1749271	0.2
2002	13	1	0.05	0.1	0.07695	0.1249479	0.2
2003	13	0.5	0.05	0.1	0.069255	0.08924853	02
2004	14	0.5	0	0.1	0.0623295	0.08924853	01
2005	15	0	0	0.1	0.059049	0.08924853	0
2006	15	0	0	0	0.059049	0.04462426	0
2007	15	0	0	0.1	0.059049	0	0
2008	16	0	0	0.1	0.059049	0	0
2009	16	0	0	0.1	0.059049	0	0
2010	16	0	0	0.1	0.059049	0	0
2011	17	0	0	0.1	0.059049	0	0
2012	17	0	0	0.1	0.059049	01	0
2013	17	0	0	0.1	0.059049	0	0
2014	15	0	0	0.1	0.0295245	0	0
2015	16	0	0	0.1	0	0	0
2016	16	0	0	0.1	0	0	0
2017	16	0	0	0.1	0	0	0
2018	17	0	0	0.1	0	0	0
2019	17	0	0	0.1	0	0	0
2020	18	0	0	0.1	0	0	0
2021	18	0	0	0.1	0	0	0
2022	22	0	0	0.1	0	0	0
2023	22	0	0	0.1	0	0	0
2024	22	0	0	0.1	0	0	0
2025	22	0	0	0.1	0	0	0
2026	22	0	0	0.1	0	0	0
2027	22	0	0	0.1	0	0	0
2028	22	0	0	0.1	0	0	0

as currently written, calls for renegotiation of payments after the first, second, and ninth years of the project. Given these payment milestones, economists in the meeting deemed it appropriate to establish a separate expansion distribution for the first year, the second year, the third through ninth years, and the remaining years.

Figure 7.3 shows a plot of the deterministic annual estimates for payments. The four sets of expansion-factor coefficients that will be used to generate expansion-factor distributions are listed in Table 7.3. Figure 7.4 depicts the annual payments in expanded form with one annual payment range shown as a distribution.

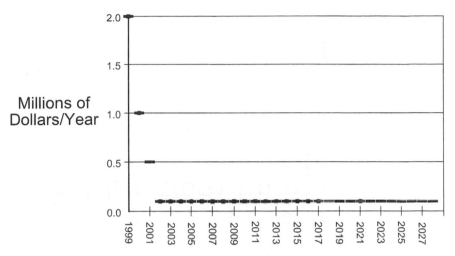

FIGURE 7.3 Time series plot of deterministic annual payments (OPES) in production sharing opportunity.

TABLE 7.3
Financial Distributions and Chances of Failure

Variable	Minimum	Most Likely	Maximum	Peakedness	Chance of Failure
Expansion Fact. — Oil Price (%)	0.9	1	1.2	5	
Expansion Fact. — Const. Cost (%)	0.8	1	1.3	8	
Expansion Fact. — Conting. Cost (%)	0.9	1	1.2	4	
Tangible Fraction CAPEX for Taxes (Integer %)		60		10	
Expansion Fact. — 1999 Payments (%)	0.95	1	1.05	5	
Expansion Fact. — 2000 Payments (%)	0.9	1	1.1	3	
Expansion Fact. — 2001-2007 Payments (%)	0.8	1	1.3	2	
Expansion Fact. — 2008-2018 Payments (%)	0.6	1	1.7	2	
Shut Down Costs ($Millions)		0.25		10	
Gov. Fails to Award Us Contract (Integer %)					15
New Party Drops Project (Integer %)					10
Probability of Regulatory Delay (%)	0.1	0.15	0.25	5	
Time Consumed by Reg. Delay (Days)	20	50	100	5	
Cost of Regulatory Delay ($Millions/Day)	0.1	0.15	0.3	5	
Corporate Tax Rate (%)		0.5		10	
Discount Rate (Integer %)		11		10	

Having cleared the time-series and expansion-factor hurdles, Hershel realizes that a good portion of those parameters that are not treated as time-series variables will be represented by single distributions that are not time dependent. After a second

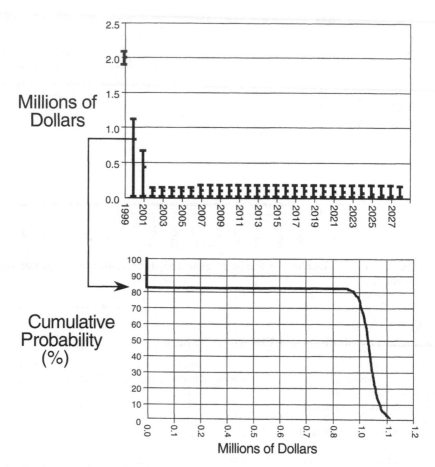

FIGURE 7.4 Time series plot of expanded payments and a distribution of payments for a single year.

day of discussion and debate, the group agrees upon the distribution-building values listed in Table 7.3. This completes, for now, the data-gathering phase. Next, of course, is the task of executing the model and interpreting the results.

RESULTS FROM THE MODEL

Hershel views the huge volume of data and plots that result from the model. It becomes immediately clear that his first job will be to select only those plots and data that are absolutely necessary to present the project case to management and partners. Other plots and data will be included in an appendix to the report. *Short, sweet,* and *clear* are the operative terms.

After perusal of all data, Hershel decides to first include in the main body of the report the oil-in-place and recoverable-oil plots shown in Figures 7.5 and 7.6,

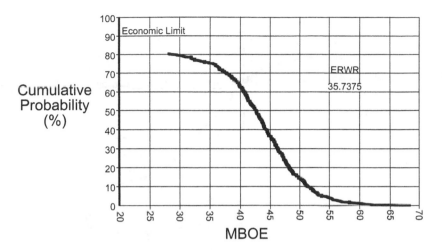

FIGURE 7.5 Cumulative frequency plot of millions BBLs of hydrocarbons in place in production sharing opportunity.

respectively. The economic limit on these plots (20 million barrels) is different from that listed in Table 7.1. In the table, production engineers determined an economic limit in units of barrels of oil per day (BOPD). This is a daily production rate below which, from their perspective, the well is no longer economical to operate. This value is used in the calculation of the Initial decline parameter.

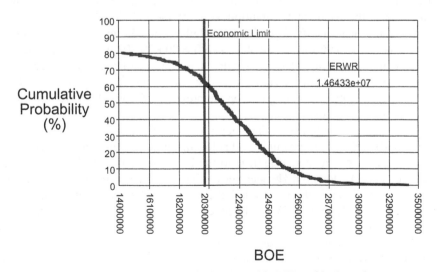

FIGURE 7.6 Cumulative frequency plot of recoverable BBLs of hydrocarbons.

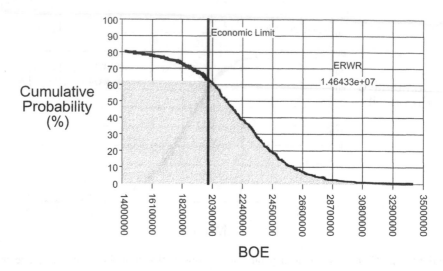

Cumulative Probability (%)

BOE

FIGURE 7.7 Cumulative frequency plot of recoverable BLs of hydrocarbons showing (hatchured) area under the curve used to determine the economic-risk-weighted-resource (ERWR) value.

The economic limit (20 million barrels) shown in Figures 7.5 and 7.6 is different. This economic limit is calculated by economists on the project and is initially calculated in dollars and then converted to barrels of oil for plotting. This value reflects all financial aspects of the full-cycle economics of the project including working interest, partnership and governmental obligations, and other considerations.

Data plotted in Figure 7.5 show the range of resources (oil) that are calculated to be in the underground reservoir. This typically is referred to as "oil in place" (OIP). Note that the chance of abject failure shown in Figures 7.5 and 7.6 is about 20%. This is the result of the multiplication of the individual chances of abject failure for the Seal quality, Trap definition failure, and Reservoir presence failure parameters ($0.9 \times 0.96 \times 0.95 = 0.8037 \times 100 = 80.87\%$ Total Chance of Success). Later, when financial variables are considered, the failure related to the government's not awarding the company the contract (15% chance) and the failure associated with the newly elected party deciding to drop the project (10%) will be considered in our Total Chance of Success calculation.

In Figure 7.5 which depicts the amount of oil in the underground reservoir, the 20 million-barrel economic limit is well below the minimum amount of oil contained in the trap (i.e., the economic limit is to the left of the left end of the cumulative frequency curve). However, Hershel knows that no oil company can recover all of the oil in an underground reservoir as indicated by the Recovery factor coefficients (percentages) in Table 7.1. When Recovery factor is applied, the resulting plot (Figure 7.6) shows the amount of oil that can actually be brought to the surface and sold. In Figure 7.6, the 20 million-barrel economic limit is well within the range of possible reserve volumes. This indicates that in addition to the (approximately) 20% chance of abject failure, there is an additional chance, even if the well does not fail

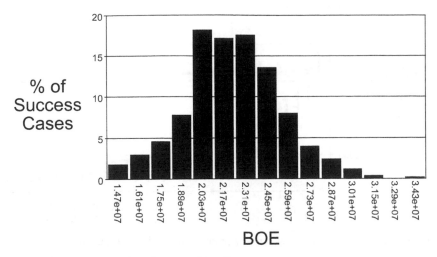

FIGURE 7.8 Frequency plot of recoverable oil.

outright (abjectly), that the well will fail financially by producing a total volume of recoverable oil less than our 20 million barrel lower limit.

Figure 7.7 shows the area under the cumulative frequency curve that is integrated to arrive at the economic risk weighted resource (ERWR) value. This value is NOT the amount of oil Hershel expects the prospect to produce if it does not fail abjectly. The expected value is a value near the peak of the frequency plot shown in Figure 7.8. The ERWR is the amount of resources that this prospect would contribute to a portfolio of prospects on a fully risk-weighted basis. It is also the number that should be used to compare one opportunity with another. A full treatment and explanation of risk-weighted values is given in the previous book.

Hershel decides that he will next include in his report some indications on how the volumes of oil will be produced over time. In Figures 7.9 and 7.10 are shown plots for the Primary Product Average Annual Daily (Production) Rate and the Cumulative Primary Production, respectively. Note that the middle tick mark (median of the distribution) on the right-most vertical bar in Figure 7.10 indicates a value somewhere between 20 and 25 million barrels of oil equivalent (BOE). This corresponds to the peak in the frequency plot of recoverable reserves and the steepest part of the cumulative frequency curve of recoverable reserves in Figures 7.6 and 7.7, respectively.

Figure 7.11 shows the expanded yearly oil price time series. This is used in the calculation (as can be seen from the equations) of annual revenue. A plot of yearly revenue is shown in Figure 7.12. Time-series plots for annual Construction costs, Contingency costs, CAPEX, and OPEX are show in Figures 7.13 through 7.16, respectively. Hershel includes these plots in the body of his report. Note that the OPEX values in Figure 7.16 reflect the four separate expansion factors. Also included in the body of the text is a plot representing the Total Cost of Regulatory Delays as shown in Figure 7.17. Now, to the meat of it. All of the aforementioned plots and

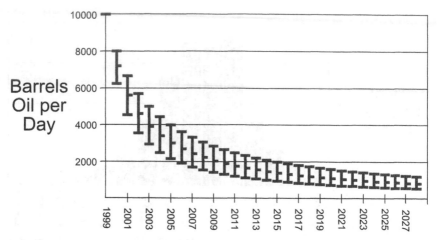

FIGURE 7.9   Time series plot of primary product annual daily production rate.

parameters culminate (see equations) in the calculation of cash flows and "summary" economic values such as net present value (NPV) and internal rate of return (IRR). A plot of annual After Tax Cash Flow is shown in Figure 7.18. Note that because of potential failure, there is always some chance of obtaining an annual cash-flow value near zero.

The cumulative frequency plots for NPV and IRR are depicted in Figures 7.19 and 7.20, respectively. Hershel notes that the total chance of failure now reflects the failure associated with the situation in which the company is not awarded the contact (15%) and the chance that the newly elected party will drop the project before it can effectively get started (10%).

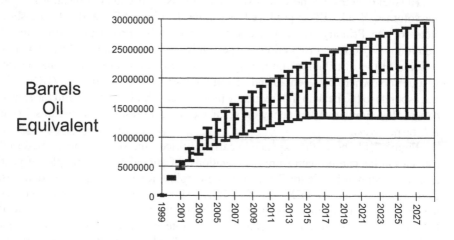

FIGURE 7.10   Time series plot of cumulative primary production.

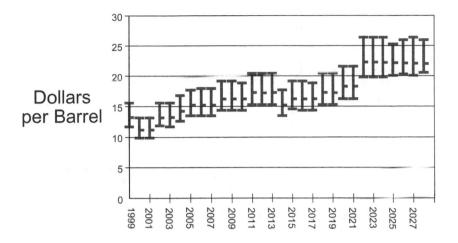

FIGURE 7.11 Time series plot of expanded oil prices.

As shown in Figure 7.19, there is some chance, if the project does not fail completely, that a negative NPV will be generated (see Figure 7.21 for a frequency-plot equivalent of Figure 7.19). There exists about a 50% chance that the project will generate a positive NPV. The positive values will have a range as shown in the frequency plot in Figure 7.21. The IRR plot shown in Figure 7.20 is similar in that it shows some chance of generating a negative IRR. There is about a 50% chance of generating a positive IRR value. The range of positive IRR values is best seen in the frequency plot of IRR in Figure 7.22. All other plots and depreciation, secondary

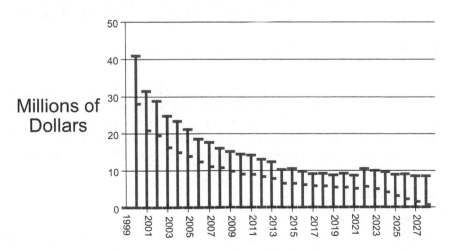

FIGURE 7.12 Time series plot of yearly revenue.

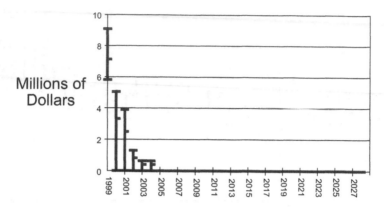

FIGURE 7.13 Time series plot of expanded construction costs.

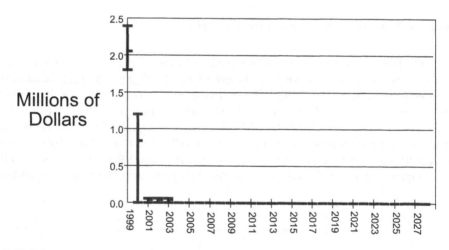

FIGURE 7.14 Time series plot of expanded contingency costs.

product production, and other items not shown here Hershel includes in the appendix of his report. The summary plots described above will be presented to management and will form the basis of their decision concern whether to proceed with this project. Given the Total Chance of Success and the chance of generating positive NPV and IRR values in the indicated ranges, would you give your OK to this project?

SO, SO WHAT?

As driven home in the first few chapters of the first book, the problem of deciding just what the question is can be by far the most time-consuming aspect of developing a risk/uncertainty model. As with the example developed in this chapter, many divergent individuals and groups might have to contribute to the delineation of the

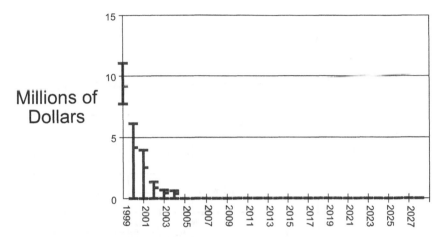

FIGURE 7.15 Time series plot of CAPEX.

problem. The process of defining the question and engaging in conversations regarding definition of the problem brings focus to the problem and helps to ensure that everyone involved is "on the same page," so to speak. Such conversations can be as enlightening (or more so) than the output from any risk/uncertainty model.

This model purposely developed in two parts. Most real business models have a significant technical aspect as well as a salient business or financial facet. Each section can be incredibly complex, but the modeling approach demonstrated in this chapter ensures that the two (or as many as needed) sections of the model are joined seamlessly. In addition, aspects such as chances of failure in the technical part of

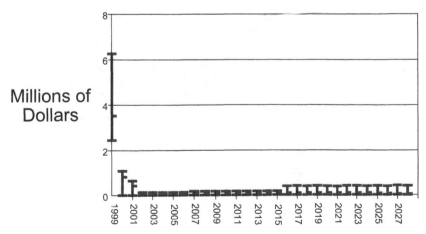

FIGURE 7.16 Time series plot of expanded OPEX (payments + regulatory delay costs + shut-in costs).

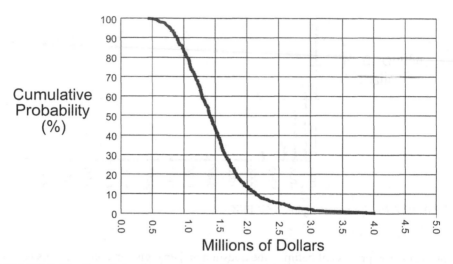

FIGURE 7.17 Cumulative frequency plot of regulatory-delay costs.

the model are brought to bear on the financial model-output parameters. This is a powerful modeling process and tool.

Validation of this model is no different from validation of other models discussed in this book. The reader is directed to the Introduction section and to the "So, So What?" discussion of validation in Chapter 4.

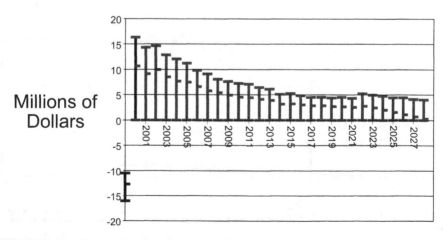

FIGURE 7.18 Time series plot of after-tax cash flow.

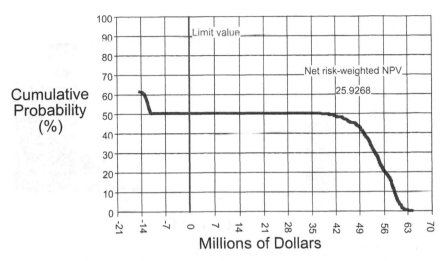

FIGURE 7.19 Cumulative frequency plot of net present value.

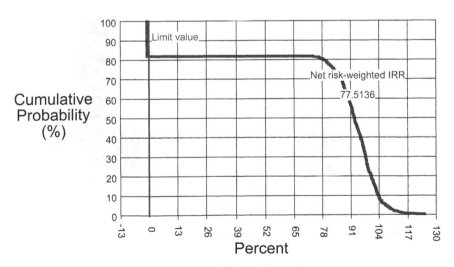

FIGURE 7.20 Cumulative frequency plot of internal rate of return.

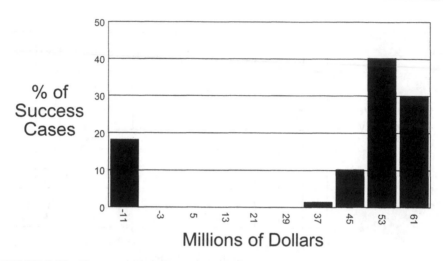

FIGURE 7.21 Frequency plot of net present value.

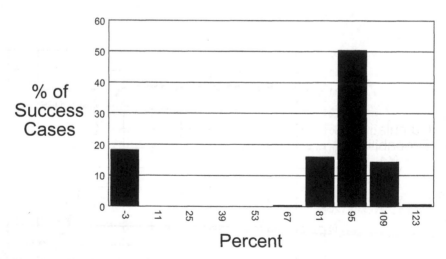

FIGURE 7.22 Frequency plot of internal rate of return.

8 Scheduling and Optimization Risk Model

CONTENTS

INTRODUCTION

Scheduling is a salient aspect of nearly every project. Scheduling of time, personnel, expenditures, construction phases, meetings, and other business-related tasks is typical. There are sophisticated software packages in the marketplace that do an admirable job of addressing the scheduling issue. In fact, a significant part of my business deals with adding a probabilistic dimension to these commercial packages (i.e., hooking our probabilistic modeling package to commercial scheduling systems).

It has been our experience, however, that most commercially available scheduling software packages assume or enforce a prescribed scheduling philosophy. Many such packages tend to be very linear in their approach (i.e., first this happens, then that happens...). Even when these products are linked with a probabilistic engine that allows iterative solution of the scheduling problem using probabilities and distributions, philosophical and system constraints can preclude performing the type of special scheduling scheme that a given problem seems to demand. Therefore, we often find that, in spite of the spate of scheduling packages available, we still have cause to construct a probabilistic model to address a scheduling problem.

Often, as in the example detailed in this chapter, the problem is not only one of scheduling, but also one of optimization. That is, practitioners are attempting to devise the *best* schedule possible. Again, there are countless packages to address the optimization issue. In general, however, these packages can suffer from one or more of three primary shortcomings. First, most of the packages are deterministic; that is, they do not allow for an iterative or probabilistic assessment of the problem. This can be partially resolved by linking these systems to probabilistic engines, but this often is cumbersome and impractical. Second, many of these systems are built for the person who is familiar with the art and science of operations research. The average person not schooled in the ways of operations techniques generally is not comfortable dealing with such systems. Third, and most vexing, is the problem that many optimization techniques can result in optimization plans that are unrealistic for the real world. That is, the optimum solutions depend upon combinations and sequences of steps that for cultural, legal, political, or other reasons could not be enacted. Getting an optimization package to deliver an optimal solution that has considered all such "soft" issues is a tall order.

So, we sometimes find it necessary and most expedient to generate a scheduling and optimization model that

- Is easy to use
- Is probabilistic
- Puts all variables under user control
- Addresses the special considerations of a problem
- Generates an optimal solution that could be put into practice in the real world

The scheduling and optimization model presented here is very simple. It is, however, an example of a risk-optimization model that is practical to implement and that can be run and understood by laypeople. The simple and practical design does present some drawbacks. First, the model presented does not iterate through a series of scenarios (it does, however, iterate within a single scenario). That is, the model leaves it up to the user to define a given situation (i.e., sets of coefficients for parameters) and makes no attempt to cycle through multiple scenarios in an attempt to find the optimal situation. This feature can be both a boon and a limitation. It is limiting because it depends on the user to manually set up and investigate various scenarios. It can be a blessing because it does not present impractical situations as optimal solutions. Certainly, the model could be easily modified to cycle through multiple scenarios in an attempt to identify the most efficient combination of parameter coefficients, but it would be counterproductive to present such a model because such models can be highly convoluted in their logic and, therefore, make terrible examples.

The fact that the model considers just one scenario at a time means that the user must enact multiple executions of the model and must manually compare the results from the independent executions. This can be time consuming. It also can be enlightening because it forces the user to think through each scenario. This, in general, ensures that only practical real-world situations are tested and time is not wasted investigating trials that could not be put into practice. This also means that there might exist an optimum solution to the problem that is simply not foreseen and tested by the user.

It is my opinion that there is, to date, no perfect solution to the scheduling/optimization problem. The process and model presented here is far from perfect and could be considered the preferred solution for only a certain class of problems. See the "So, So What?" section of this chapter for more discussion.

THE PROBLEM

Ray is the production manager at a fabrics plant. The facility is about to begin producing a new line of polyester fabric. Raw materials for the new fabric will arrive in liquid form at the plant site. Tanker-type rail cars will transport the liquid raw material from the supplier's facility to Ray's fabric plant. New storage tanks with a maximum capacity of 200,000 (200K) pounds have been constructed on the fabrics plant site. These tanks will store the liquid raw material.

As production manager, just one of Ray's concerns is balancing the flow of raw materials into the plant with the consumption of the raw materials for fabric production. Of course, Ray would like to run out of raw materials as infrequently as possible. The plant is scheduled to run 7 days a week, so running out of raw materials necessarily means shutting down the plant which, in turn, means lost revenue.

Storage capacity for raw materials is limited, as previously mentioned, to 200K pounds. The plant has only limited capacity to "store" train cars at the railhead of the factory rail spur. Due to this limited capacity and because economics dictate that train cars be processed as quickly as possible, train cars carrying raw material must be rolled up to the tanks, unloaded, and removed from the plant site in the most expeditious manner.

In addition to the storage-capacity concerns, Ray's task of scheduling raw materials delivery is complicated by several other factors. First, the railroad and raw-materials supplier are companies with which the fabrics plant already has contracts for other product lines. Ray knows that it takes 1 to 3 days to get a rail car from the supplier's site to his fabrics plant. Due to weather, switching-yard complexities, and other considerations, the probability of a car (or set of simultaneously sent cars) making the trip in 1, 2, or 3 days is equal.

Second, Ray's plant runs 7 days a week. The raw materials supplier's plant does not; no rail cars are shipped from the supplier's site on Saturday and Sunday. In addition, unionized railroad workers needed to transport the rail cars along the privately owned fabrics-plant rail spur do not work on weekends. Therefore, train cars can be in transit on main rail lines during weekends, but if they arrive at the fabrics plant on Saturday or Sunday, the cars must be "parked" at the railhead to be transported to the plant on Monday morning.

There exists at least one more major complication. Because of start-up problems, first-year consumption of raw materials at the plant is expected to be erratic. Ray's engineers have informed him that he can expect average daily raw materials consumption to vary from 20K to 60K pounds with a most likely consumption of about 40K pounds per day.

There are, however, a couple of things Ray can count on. One is the capacity of the rail cars. For safety reasons, the supplier will not ship a partially filled liquid-tank car. Ray knows, then, that each car will contain 10K pounds of raw materials. Ray also knows that he will start his plant with an initial inventory of 100K pounds of raw materials left over from a previous fabrics plant process.

At first, Ray tries to juggle all of these considerations in his head, but the convoluted interrelationships between all the variables begin to be too much for him to handle. It was time, he reasoned, to seek help.

Sigfred is the local statistician and risk/uncertainty expert. Following an initial meeting, Sigfred and Ray agree to attempt to construct a simple probabilistic model to help Ray determine an efficient schedule for the shipment of rail cars.

DESIGN OF THE MODEL AND THE CONTRIBUTING-FACTOR DIAGRAM

At a second meeting, the two plant employees begin to map out their strategy. The duo first decides to generate a list of variables that will constitute the model. Next, they will decide on reasonable coefficients for each parameter. These parameters will then be linked in a contributing-factor diagram. The first step will be to use the contributing-factor diagram as a blueprint to construct and populate a probabilistic model.

Because time is short and because the scheduling dilemma is but one of Ray's many charges, the two decide to build a model of relatively high level — that is, a model not too burdened with detail. Their initial list of input variables for the model is

- Transit time for rail cars
- Number of rail cars sent daily from the supplier
- Car capacity
- Fabrics-plant daily consumption of raw materials
- Initial inventory of raw materials
- Starting day of the week for the model

In the risk model, all of these parameters will be incorporated as true variables with the ability to be represented as single values, distributions, or time series. The relatively simple contributing-factor diagram linking the model elements is shown in Figure 8.1.

Table 8.1 lists all of the non-time-series distribution coefficients. As can be seen in Table 8.1, Sigfred and Ray know that the capacity of train cars used is 10K pounds. Safety concerns preclude the transport of partially filled cars, so each car that arrives at the fabrics plant yields 10K pounds of raw material.

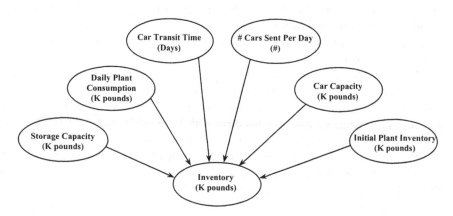

FIGURE 8.1 Contributing-factor diagram for scheduling risk model.

TABLE 8.1
Non-Time-Series Scheduling-Model Variables

Variable	Minimum	Most Likely	Maximum			
Car Load (K Pounds)		10				
Daily Plant Consumption (K Pounds)	20	40	60			
Initial Plant Inventory (K Pounds)		100				
Starting Day (Number 1–7)		3				
	Value	Weight	Value	Weight	Value	Weight
Car Transit Time (Days and Percent)	1	33.333	2	33.333	3	33.333

A second spike (single-valued) variable is the number of pounds of raw material already on hand at the fabrics plant. Ray knows that there exist 100K pounds of required raw material already on the plant site. This material now occupies half of the fixed 200K pound storage capacity at the plant.

The risk model considers a time series of days. All days are not equal. The fabrics plant operates over the weekend but because of railroad labor constraints, rail cars cannot be run along the 6-mile fabrics-plant rail spur to the storage tanks. On weekends, rail cars are "parked" at the head of the rail spur and then delivered on Monday morning. Therefore, it is critical that the model "know" on what day of the week (1 through 7) the model will commence. A starting-day variable represented by a single value (1 to 7) will be used to indicate the day of the week on which the model is to begin.

Two model variables are represented by distributions. As previously described, the transit time is one to three days for a single car or for a group of rail cars that are, effectively, simultaneously sent from the raw materials supplier. Experience has shown that the probability of cars making the trip in 1, 2, or 3 days is equal. This transit-time variable is represented as a discrete distribution. A graphic depiction of this distribution is shown in Figure 8.2.

Fabrics-plant consumption of raw materials will be variable. Because of start-up problems, unpredictable capacities and efficiencies of plant components, and other considerations, best-guess estimates indicate that daily raw-material consumption will be between 20K pounds and 60K pounds with a most likely consumption of 40K pounds. This distribution is shown in Figure 8.3.

A time series is used to represent the final variable in the model. Ray's ultimate goal is to schedule the shipment of train cars from the sending location so that the fabrics plant is unlikely to either run out of raw materials or exceed its 200K pounds of storage capacity. Many considerations complicate this task. Train cars are sent from the raw-materials supplier 5 days a week. The fabrics plant consumes resources 7 days a week. Raw material, however, is not put into the storage tanks on weekends (train cars parked at railhead) and a "flood" of raw materials descends on the storage tanks on Monday morning.

On any given day, the amount of raw materials arriving at the fabrics plant storage tanks depends partly on the transit time (1 to 3 days) for rail cars, partly on the day of the week it is, but mainly on how many train cars are actually sent on a given day

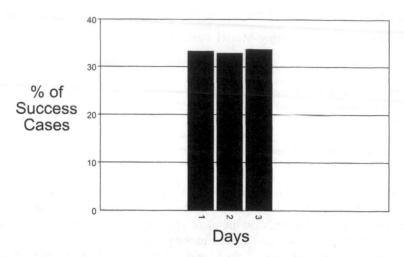

FIGURE 8.2   Frequency plot of discrete distribution for car transit time.

from the raw materials supply location. The supplier insists they can guarantee shipment of up to six rail cars a day. A maximum of 10 cars can be achieved for, at the most, two consecutive days. Supply and logistical constraints at the supply site preclude reaching the maximum level of ten cars more than 2 days in a week.

Not being able to juggle all of these variables in his head, Ray decides to assign to the single-valued and distribution-based variables the coefficients shown in Table 8.1. Ray does not have an intuitive feel for the effect of varying the number of train cars sent each day. He decides that for the initial trial run of the risk model, he will

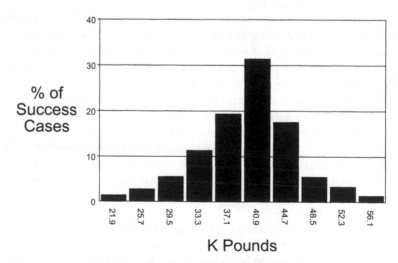

FIGURE 8.3   Frequency plot of daily raw-materials consumption.

TABLE 8.2
Time Series Variable for Scheduling Model

Day	Number of Cars Sent
Wednesday week 1	6
Thursday week 1	6
Friday week1	6
Saturday week 1	0
Sunday week 1	0
Monday week 2	6
Tuesday week 2	6
Wednesday week 2	6
Thursday week 2	6
Friday week 2	6
Saturday week 2	0
Sunday week 2	0
Monday week 3	6
Tuesday week 3	6
Wednesday week 3	6
Thursday week 3	6
Friday week 3	6
Saturday week 3	0
Sunday week 3	0
Monday week 4	6

---

indicate in the time series variable that the same number of cars will be sent each day. These time series coefficients are listed in Table 8.2. Before Ray can run the risk model, however, it must be built. Ray convenes yet another meeting with Sigfred to map out the logic for the model.

THE RISK-MODEL CODE

Following discussions of all complicating factors, Sigfred announces, much to Ray's relief, that the program for the model will be relatively simple to construct. The computer logic will begin with some mundane "housekeeping" chores. Among the initial steps will be

- Establishing numerical values, ranging from 0 to 6, to determine how many days after the model's initial day is Saturday
- Initializing variables
- Determining the volume of raw materials that arrives at the fabrics plant rail head each day

Following this section of the intuitive calculations, a more convoluted section ensues. In this latter "while loop" is determined the amount of inventory on hand in storage tanks at the end of any given day. Because inventory is limited to a maximum of 200K pounds and because complete depletion of inventory would shut down the plant, the calculation of the day-ending inventory volume is the crux of the problem.

This section of the model is complicated, for example, by the fact that train cars can travel to and arrive at the fabrics plant railhead on Saturdays and Sundays, but cannot contribute to the inventory volumes in the storage tanks. Because the fabrics plant continues to operate on weekends, inventory will be depleted without replenishment until Monday morning. On Monday morning "comes the flood." Train cars that arrived on Saturday and Sunday are brought to the storage tanks for unloading. Inventory volumes on Monday need to be sufficiently depleted for the storage tanks to be able to accommodate the pent-up volumes.

Another complication looms on Tuesday and Wednesday. As previously stated, the transit times for cars is 1 to 3 days. No cars are sent from the supplier's facility on the weekend. Therefore, unless Ray compensates for this in the model, he could expect reduced car arrivals and inventory on Tuesdays and Wednesdays (Mondays too, but the "parked cars" will alleviate this problem on Mondays).

The computer code generated by Sigfred is shown below.

```
inventory = initial;
if start_day = 1 then sat = 5; endif
if start_day = 2 then sat = 4; endif
if start_day = 3 then sat = 3; endif
if start_day = 4 then sat = 2; endif
if start_day = 5 then sat = 1; endif
if start_day = 6 then sat = 0; endif
if start_day = 7 then sat = 6; endif
per = 0;
while per < _NUM_PER do
    inventory[per] = 0;
    arrivals[per] = 0;
    per = per + 1;
enddo
shipdate = 0;
while shipdate < _NUM_PER do
    arrivedate = shipdate + trans_time;
    if (arrivedate < _NUM_PER) then
        arrivals[arrivedate] = arrivals[arrivedate] + cars_per_day[shipdate] *
            car_load;
    endif
    shipdate = shipdate + 1;
enddo
per = 0;
inventory[per] = initial;
per = 1;
while per < _NUM_PER do
    start_inv = if(inventory[per − 1] > 0, inventory[per − 1],0);
    if per%7 = sat then
```

```
        inventory[per] = start_inv – consumption;
    elseif (per - 1)%7 = sat then
        inventory[per] = start_inv – consumption;
    elseif (per = 1) then
        inventory[per] = start_inv + arrivals[per] – consumption;
    elseif (pcr – 2)%7 = sat then
        inventory[per]  =  start_inv  +  arrivals[per-2]  +  arrivals[per-1]  +
    arrivals[per] – consumption;
    else
        inventory[per] = start_inv + arrivals[per] – consumption;
    endif
    per = per + 1;
enddo
```

where

inventory is the volume of raw materials in the storage tanks
initial is the beginning inventory
start_day is the first day considered by the model
sat is a variable that holds a value ranging from 0 to 6 that represents the
    number of days between the starting day and Saturday
per is a counter for the number of periods (days)
_NUM_PER is a variable that holds the number of time series periods for
    the model
arrivals is the number of train cars that arrive at the fabrics plant on a given day
shipdate is the day of the week a car or set of cars leaves the sending location
arrivedate is the day of the week on which train cars arrive at the fabrics plant
trans_time is the amount of transfer time
cars_per_day is the number of cars sent from the supply location on a given day
car_load is the amount of raw material carried by a train car
start_inv is the amount of inventory on hand at the beginning of a given day
consumption is the amount of inventory consumed in a given day
% is a function that returns the remainder of a division (modulus function)

RESULTS FROM MODEL EXECUTION

The time series inventory plot resulting from using the coefficients in Tables 8.1 and 8.2 is shown in Figure 8.4. Figure 8.5 is a plot of the inventory distribution for the first Sunday in Figure 8.4. This plot clearly shows a significant chance (about 58%) of depleting the inventory. Subsequent Sundays also exhibit unacceptable probabilities of running an inventory deficit.

Mondays also are a problem. Figures 8.6 and 8.7 are cumulative frequency plots for second and third Mondays, respectively. Both plots indicate that there is an (average) 20% risk of exceeding the 200K pound storage limit on Mondays due to the buildup of inventory over the weekend.

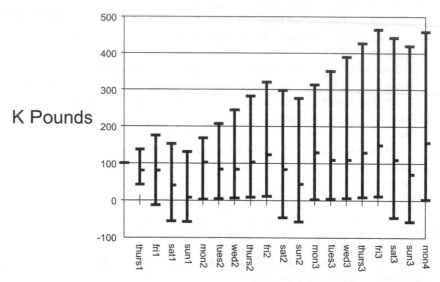

FIGURE 8.4 Time series plot of plant inventory levels.

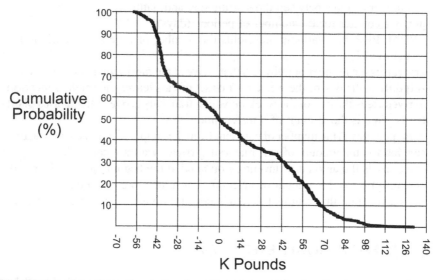

FIGURE 8.5 Cumulative frequency plot of inventory on the first Sunday.

To attempt to ameliorate this dilemma, Ray believes the best first step is to hold all other variables at the single values and ranges shown in Table 8.1 and to experiment with the time series coefficients in Table 8.2. This can be done only if the variable represented by the coefficients in Table 8.2 is truly independent from other variables — and it is. If he can approach a reasonable solution, Ray might then attempt to fine-tune the model by experimenting with coefficients for other variables in the model.

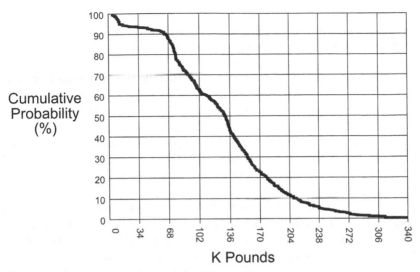

FIGURE 8.6 Cumulative frequency plot of inventory on the second Monday.

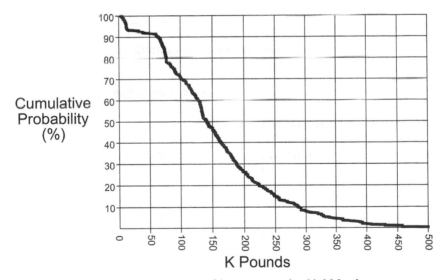

FIGURE 8.7 Cumulative frequency plot of inventory on the third Monday.

After thinking through the many complications and following many runs of the model using various time-series scenarios, Ray decides that the time series of number of cars shipped shown in Table 8.3 does a fair job of balancing inventory shortfalls and overruns. Ray has come to the conclusion that he is going to have to live with the fact that on Sundays he might have to shut down the plant for at least part of the day because of shortfall in inventory. Mondays will always harbor the specter of having more inventory on hand than can be stored. Fridays also pose a problem due to the inventory that must be built up for operations on the weekend.

Figure 8.8 is the time-series plot resulting from using the time-series coefficients shown in Table 8.3. As can be seen from cumulative frequency plots of first and second Sunday inventories (Figures 8.9 and 8.10, respectively), Sundays still pose a threat of inventory shortfall. However, the probability of exhausting inventory on Sunday is diminished on subsequent Sundays. Ray is pleased that he has succeeded in attaining mean inventory values (middle tick marks on vertical bars in the time series plot) near 100K points on most days. Later days in the model show significant upsides in inventory, but as can be seen from the cumulative frequency plot of inventory for the final day, the probability of exceeding 200K pounds is acceptably small. The increase in "upside" inventory with passing time, Ray reasons, will have to be attacked by attempting to push production to upper limits on Mondays. This can be done economically by increasing the labor force and shift lengths on Mondays.

Ray is very pleased with the results of the analysis. The model has allowed him to approach the problem in a probabilistic manner.It also has facilitated experimentation with practical real-life scenarios one (or more) variable at a time. All of the complex relationships were succinctly handled in a very short period of time. Ray will apply this probabilistic scheduling approach on other fabrics-plant problems.

TABLE 8.3
Time Series Variable For Optimized Scheduling Model

Day	Number of Cars Sent
Wednesday week 1	10
Thursday week 1	9
Friday week 1	5
Saturday week 1	0
Sunday week 1	0
Monday week 2	3
Tuesday week 2	3
Wednesday week 2	9
Thursday week 2	7
Friday week 2	5
Saturday week 2	0
Sunday week 2	0
Monday week 3	3
Tuesday week 3	3
Wednesday week 3	9
Thursday week 3	7
Friday week 3	4
Saturday week 3	0
Sunday week 3	0
Monday week 4	3

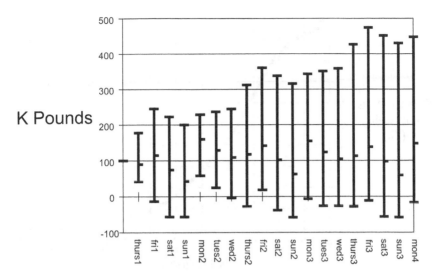

FIGURE 8.8 Time series plot of inventory.

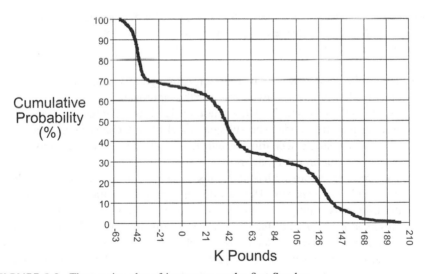

FIGURE 8.9 Time series plot of inventory on the first Sunday.

SO, SO WHAT?

Optimization models such as the one presented here are not necessarily superior in all respects to commercially available scheduling or scheduling/optimization packages. Every approach has its good points and its drawbacks. For different perspectives on optimization techniques applied to portfolios and other problems, see books by Schrijver, Rustem, Goldberg, Haupt, Malkiel, and Markowitz in the Selected Readings section of this book. Linear programming techniques, genetic algorithms,

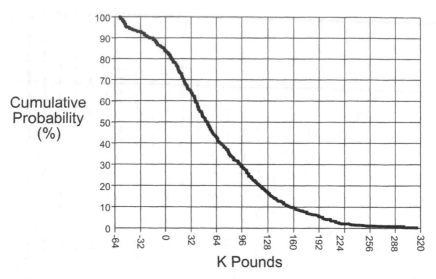

FIGURE 8.10   Time series plot of inventory on the second Sunday.

efficient frontier methods, and many other diverse technologies are applied in many situations to attempt to achieve an optimal solution. The following are the primary drawbacks to this approach.

- The model must be written by the user. Although this is also true for many other packages, it certainly is true for this approach.
- The model presented here does not cycle through multiple scenarios seeking the "best" one. The subject model of this chapter considers all aspects of a single scenario through the iterative process, but considers only one scenario per model run.
- The presented approach has no means, as constructed, to compare results from multiple scenarios. This is different from the point above. Even if the user runs multiple scenarios in subsequent model runs, the model presented here does not facilitate the comparison of multiple-model-run-output variables to help determine which scenario is the optimal solution.
- The truly optimal solution to the problem might not be discovered because the user did not think to test the most efficient scenario. The model does not facilitate discovery of that situation.

Conversely, the primary benefits to this modeling technique are as follows.

- It is easy to use — even for those who are not experts in the fields of modeling or optimization. This is a huge advantage in the real business world.
- No unrealistic scenarios are tested or presented as possible solutions. Only user-prescribed situations are modeled.

- Scenario selection and design are under the user's control. Practical experience and judgment are easily incorporated.
- How the results were arrived at can be explained. This is a major advantage, especially when model results have to be presented to people not directly attached to the project.

Because I designed the example and model presented in this chapter, I could have had it turn out any way I liked. Note that the final scheduling solution presented here does not result in a perfect material balance at the plant. This is real life. Tradeoffs and compromises almost always have to be made. In the example presented in this chapter, it is clear that unless some other changes can be made, the material balance in inventory simply cannot be optimized for each day. In order to make the inventory situation on some days better, it is necessary to negatively impact the inventory levels on other days. The solution to Ray's problems might include a suggestion to management that the physical nature of the plant be modified — more storage tanks, a modification of the railroad situation, or some other means of changing the game.

9 Decision/Option-Selection Risk Model

INTRODUCTION

Often, a decision to be made is actually a choice. That is, for a given problem, there might be several rational and reasonable solutions from which to select. In a business scenario, each option's value (NPV) to the company can be calculated. The value measure should then be used to rank and prioritize the options so that a single optimal solution can be chosen.

A given solution to a problem can itself be a complex entity. The many parameters that compose the option are generally variables — that is, a single parameter can take on a range of values. For example, an investment portfolio manager might be faced with constructing an "optimal" (with respect to return) portfolio of investments. Parameters such as purchase price, interest rates, and others certainly vary over time. Such parameters, in any model, would therefore be represented as variables. In a probabilistic model, distributions of values would be used to emulate the range of real-world values for a given parameter.

If the various solutions are similar in type, a single stochastic model can be constructed and used to assess all options. Typically, this is done one option at a time by feeding a given solution's data into the model, rerunning the model, and collecting and reviewing the output. This process is repeated for each possible solution. In the end, the probabilistic results from each model run are compared (possibly using a second stochastic model that uses as input the output files from the first model) to aid in the selection of the optimal solution.

In the narrative that follows, our subject will be faced with the problem of individually evaluating multiple options for solution to a problem. Each option is composed of multiple variables. A risk model will be proposed to probabilistically evaluate each possible solution and an optimal answer will be selected.

THE CURRENT SITUATION

Felix is Director for Supply and Logistics at Plastingcap, a global producer of plastic products. Felix has just been informed that within 2 months' a new line of products will begin production at the plant. As a result, throughput of raw materials at the plant is expected to increase dramatically. As the head of shipping, Felix is responsible for ensuring that the plant has the raw-materials supply it needs when it needs it.

Preliminary estimates from engineers indicate that the increase in raw-materials consumption is expected to be as low as 50% or as high as 75%. Currently, the plant is supplied with raw materials by a barge that makes runs between the raw-materials supplier and Felix's plant. The barge currently is running at approximately maximum capacity. Any increase in production, especially one as dramatic as is forecast, will require a change in the conveyance used to deliver raw materials.

Depending on river conditions and weather, the currently employed barge can carry between 35 and 42 tons of raw materials in a single trip. Such a round trip is 10 to 15 days in duration. Current raw-materials consumption at the plant is, under ideal conditions, 2 to 4 tons per day. Distributions for these three parameters are shown in Figures 9.1 through 9.3, respectively.

Felix has traditionally used a very simple stochastic model to estimate delivery volumes and plant consumption. The simple model calculates the barge delivery capacity over any period of time and calculates the balance between supply and consumption. For a 30-day period, the barge delivery capacity and material balance at the plant are calculated in the simple risk model by the equations:

BargeCapPerPerd = CurBargeCap * (Period/CurBargeTime);
CurBalance = BargeCapPerPerd – (Period * CurConsumption);

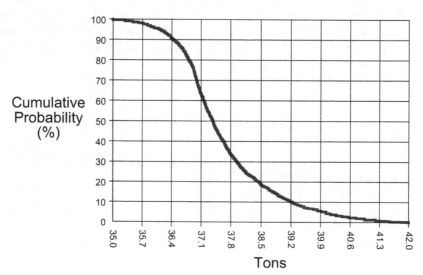

FIGURE 9.1 Cumulative frequency plot of current-barge capacity.

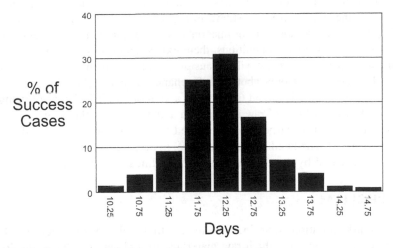

FIGURE 9.2 Frequency plot of current-barge round-trip time.

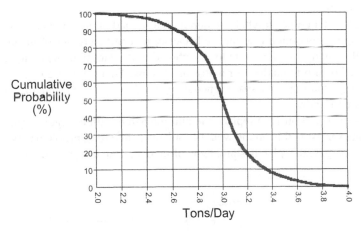

FIGURE 9.3 Cumulative frequency plot of current plant daily raw-materials consumption.

where:

   BargeCapPerPerd is the raw-material-delivery capacity of the barge over the time period of the model
   CurBargeCap is the capacity of the currently utilized barge
   Period is the number of days over which the model is run
   CurBargeTime is the time it takes the current barge to make one round trip from supplier location to the plant
   CurBalance is the raw-materials balance at the plant over the period considered
   CurConsumption is the volume of raw materials consumed at the plant over the period considered

Plots of the barge delivery capacity for a 30-day period and the plant-materials balance for the same period are shown in Figures 9.4 and 9.5, respectively.

Figure 9.5 shows the current-materials balance cumulative frequency plot. It indicates that, under ideal conditions, there exists about a 40% chance that in any given 30-day period, the plant will exhaust its supply of raw materials. The plot also shows that there exists only about a 10% chance of being short by about 2 days' volume of raw materials (about 6 to 8 tons) and about a 20% chance of being short by about 1-day's volume of raw material. In actual practice, the accumulated down time for the plant has nearly always exceeded 5 days in a 30-day period due to both scheduled and unscheduled events. Therefore, in actual practice, the shortage of raw materials indicated by the plot in Figure 9.5 is almost never realized.

THE PROBLEM

As previously mentioned, an increase in raw-materials consumption will force a change in the manner in which raw materials are delivered to the plant. Because he is in the business of shipping, Felix is well aware of what options are available to increase the capacity of raw-materials delivery. Fundamentally, he is faced with three options.

Felix's first choice might be to purchase a second barge identical to the currently used barge. This option is attractive because the characteristics of the current barge are well known. The downside to this choice is the fact that such barges cannot be connected or run in tandem for a host of reasons. Therefore, use of two barges will necessarily result in dock-space scheduling problems. These problems, however, are considered minor.

A second option is to replace the current barge with a larger barge. While it is true that exercising this option would avoid any scheduling problems associated with multiple barges, this plan has problems of its own. The barge currently in use barely

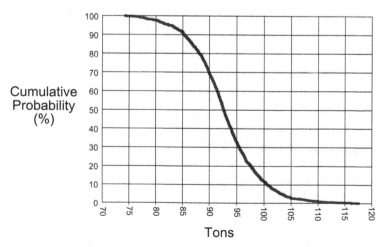

FIGURE 9.4 Cumulative frequency plot of current-barge delivery volume for 30-day period.

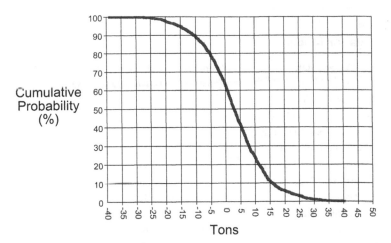

FIGURE 9.5 Cumulative frequency plot of current plant raw-materials balance for 30-day period.

fits in the locks through which it has to pass. In addition, during dry seasons, water levels in the river come close to being insufficient to allow passage of the present barge. Felix believes that a larger barge would exacerbate this situation and materials delivery to the plant could be halted altogether.

A third alternative is to replace the barge with a ship. A ship has navigation options that effectively negate the low-water problems presented by the single larger barge. However, a ship is relatively swift compared with a barge and a ship's capacity is relatively great. Felix doubts that a ship would be an efficient option because, due to its relatively great speed and capacity, it would experience a considerable amount of idle time.

Sales margins for finished products produced at the plant make the consideration of the larger crew for the ship or the second crew for the additional barge scenario insignificant. However, the taxes the plant pays on raw-materials inventory are not insignificant. Although storage capacity for raw materials is not a problem, Felix is well aware that, because of taxes and storage costs, he needs to keep on-site raw-materials volumes to a minimum.

All of this is more than Felix can cognitively juggle. Felix knows that he will need help to revamp his simple computer program to handle this more complex situation and to help him select the optimal path. Oscar, the on-site risk assessment and computer expert, is consulted and engaged.

Following several initial meetings, Oscar believes he has a grip on the problem. He proposes to Felix that he construct a risk model that will calculate the materials balance at the plant for each of the three options. Felix agrees and supplies Oscar with the data shown in Table 9.1.

Upon receipt of the data, Oscar generates the following risk-model logic.

NewConsumption = CurConsumption + (CurConsumption * ProdInc);
IdentBarge = CurBargeCap * (Period/NewBargeTime);

TABLE 9.1
Raw Data for Model

Current State	Minimum	Most Likely	Maximum	Peakedness
Barge Capacity (Tons)	35	37	42	5
Barge Trip Time (Days)	10	12	15	5
Plant Daily Consumption (Tons/Day)	2	3	4	5
Number of Days Considered (Days)		30		
Production Increase				
Plant Projected Increase (Decimal %)	0.5	0.6	0.75	
Identical Barge				
Trip Time (Days)	10	12	15	5
Large Barge				
Capacity (Tons)	50	65	70	5
Trip Time (Days)	7	8	9	5
Ship				
Capacity (Tons)	80	85	90	5
Trip Time (Days)	3	4	5	5

IdentBalance = (BargeCapPerPerd + IdentBarge) – (Period * NewConsumption);
BigBarge = BigBargeCap * (Period/BigBargeTime);
BigBalance = BigBarge – (Period * NewConsumption);
Ship = ShipCap * (Period/ShipTime);
ShipBalance = Ship – (Period * NewConsumption);

where, in addition to the model parameters previously described

NewConsumption is the projected daily consumption for the plant
ProdInc is the projected percent increase in plant production
IdentBarge is the delivery capacity of the identical barge over the considered
    time period
NewBargeTime is the round-trip time for the identical barge
IdentBalance is the plant raw-materials balance over the considered period
    for the twin-barge scenario
BigBarge is the delivery capacity of the large barge over the considered period
BigBargeCap is the raw-materials capacity of the large barge
BigBargeTime is the round-trip time for the large barge
BigBalance is the plant raw-materials balance over the considered period for
    the large-barge scenario
Ship is the delivery capacity of the ship over the considered period
ShipCap is the raw-materials capacity of the large barge
ShipTime is the round-trip time for the large barge
ShipBalance is the raw-materials balance over the considered period for the
    ship scenario

RESULTS FROM MODEL EXECUTION

Figures 9.6 through 9.11 show the frequency plots resulting from the minimum, most likely, maximum, and peakedness values for the remainder of the variables in Table 9.1. Figure 9.12 shows the calculated daily raw-materials-consumption rate at the plant. Oscar's Monte Carlo–based risk program was instructed to perform 1500 iterations. Oscar and Felix first peruse the program output related to the first option — that of adding a second identical barge.

Figure 9.13 shows a cumulative frequency plot of the identical-barge delivery capacity over a 30-day period. Even though the proposed additional barge is identical to the original craft, Felix and Oscar have decided to calculate in the risk model a

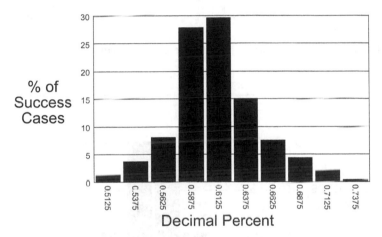

FIGURE 9.6 Frequency plot of percent projected production increase at plant.

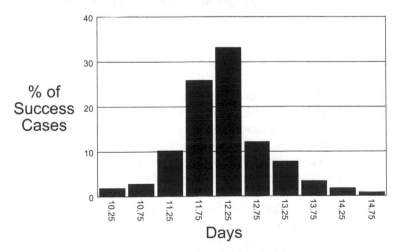

FIGURE 9.7 Frequency plot of round-trip time for identical barge.

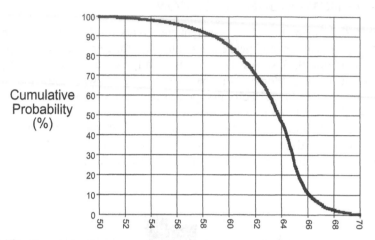

FIGURE 9.8 Cumulative frequency plot of large-barge capacity.

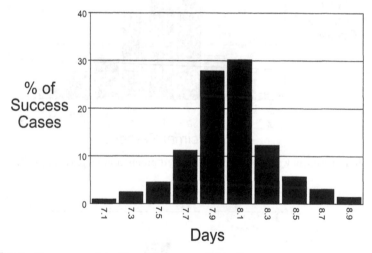

FIGURE 9.9 Frequency plot of large-barge round-trip time.

separate capacity plot for the new barge.They reasoned that this was a prudent and not redundant step because deliveries by the two barges would be staggered in time. River conditions can change rapidly and, therefore, trip times and capacities for the twin crafts might have to be independently manipulated in future runs of the model.

Figure 9.14 depicts a cumulative frequency plot of the 30-day materials balance for the twin-barge scenario. Inspection of the plot indicates that there exists only about a 2% chance of running short of raw materials. The plot also indicates that the twin-barge scenario, over a 30-day period, will result in an oversupply of raw materials. The plot indicates that there is a 50% chance of having an oversupply of about 40 tons over 30 days. This is not ideal. However, Felix knows that, over time, the demand for raw materials will increase and, thus, the magnitude of oversupply will diminish.

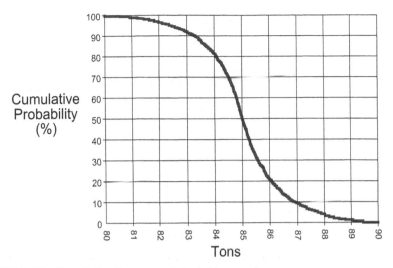

FIGURE 9.10   Cumulative frequency plot of ship capacity.

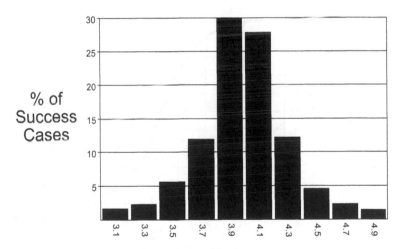

FIGURE 9.11   Frequency plot of large-barge ship time.

Figures 9.15 and 9.16 are, respectively, plots for the large-barge delivery capacity over 30 days and the plant raw-materials balance over the same time period. The materials-balance plot shows that over a 30-day period, the large barge would result in a significant oversupply of raw materials. The plot indicates that there is a 50% chance of an oversupply of more than 90 tons in a 30-day period. This is of concern for at least two reasons. First, it is important to make most efficient use of the barge and tugboat crew. It is not a real option to repeatedly employ, lay off, and then rehire the tug and barge crew in order to effect periodic halts in deliveries. The only real economic and contractual option is to keep the barge, tug, and crew in constant, or near constant, use. Therefore, enacting relatively frequent halts in deliveries to

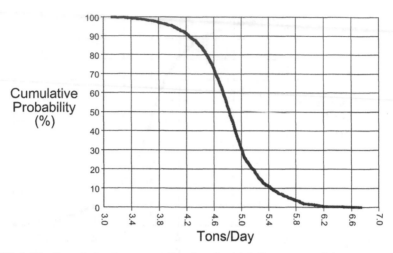

FIGURE 9.12 Cumulative frequency plot of projected daily raw-materials consumption for the plant.

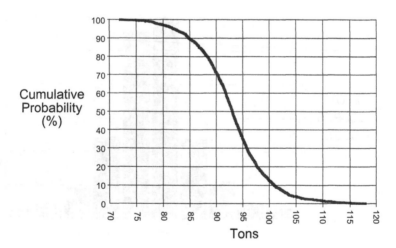

FIGURE 9.13 Cumulative frequency plot of identical-barge delivery capacity over 30-day period.

control oversupply is not a realistic option. Hauling partial loads with the barge is not an efficient use of the larger barge. In addition, less-than-capacity loads in the barge results in the barge riding dangerously high in the water, creating an unstable and unwieldy mate for the tug. This is a safety hazard.

A second reason for concern regarding large-volume oversupply concerns taxes. The plant is in part taxed by the state based on the amount of raw materials carried as inventory at the plant. Storage of large volumes of raw materials necessarily means a significant increase in taxes. This, of course, is to be avoided if possible.

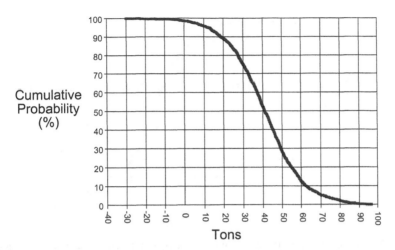

FIGURE 9.14 Cumulative frequency plot of plant raw-material for twin-barge scenario over 30-day period.

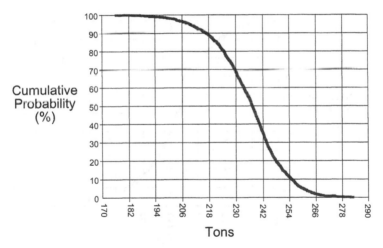

FIGURE 9.15 Cumulative frequency plot of large-barge delivery capacity over 30-day period.

So, at first glance, the larger-barge option does not appear to be as appealing as the twin-barge scenario.

The final option considered by the model is to replace the current barge with a ship. Figures 9.17 and 9.18 from Oscar's risk model show, respectively, plots for the ship's delivery capacity over the 30-day period and the plant raw-materials balance.

Scrutiny of the information in Figure 9.18 makes it clear to Felix that the ship is both too fast and is of too great a volume to be a serious contender as a raw-materials conveyance. The only ways to make the ship a practical option would be to make the ship idle for relatively long periods of time or to share the ship with

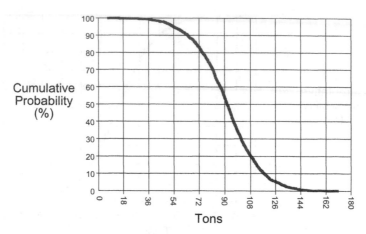

FIGURE 9.16 Cumulative frequency plot of plant raw-material balance for large-barge scenario over 30-day period.

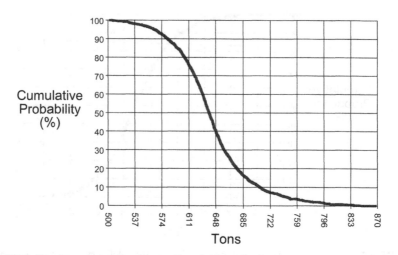

FIGURE 9.17 Cumulative frequency plot of ship capacity.

other companies. Idling the ship necessarily means idling the crew. Economics dictate that this is not an attractive alternative. Sharing the services of the ship makes more economic sense; however, Felix knows that because of erratic swings in plant production, demands for raw-materials delivery also will be erratic. Time-sharing a ship would not afford Felix the flexibility he needs in his delivery schedule.

Felix knows that the twin-barge option is not perfect. However, considering the output from Oscar's model for all three options, it affords him the delivery rates and flexibility in delivery required. Although Felix has fundamentally concluded to go with the twin-barge option, before making the final decision he may ask Oscar to

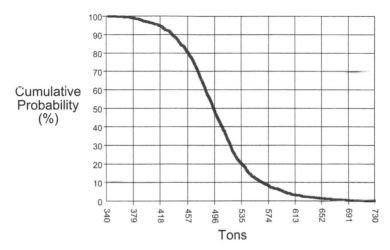

FIGURE 9.18 Cumulative frequency plot of plant raw-materials balance for ship scenario over 30-day period.

modify his model to account for the purchase and operating costs for the three options. That, however, is for another day.

SO, SO WHAT?

By design, the process and model presented here are relatively simple. This presentation option was purposely selected because most real-world option-selection models are so convoluted and complex as to make them useless as examples in a treatise such as this book. Also, although the model presented in this chapter is deemed an option-selection model, it is fundamentally different from the optimization situation outlined in Chapter 8. In that situation, the subject was attempting to design an optimal solution by manipulating the coefficients for one or more variables (cars shipped per day, etc.). In the situation presented in this chapter, all parameter coefficients are fixed (that is, ranges are fixed —for example, barge or ship capacities cannot be changed) and the number of situations from which to choose is limited to three.

In the example presented here, Felix makes his final option selection by manual inspection and comparison of the output plots associated with each of the three options. The reader should keep in mind that a model could have been constructed (as was done in the terrorism situation in Chapter 2) to compare options by algorithm.

Again, the outcome of the example has been designed to be somewhat ambiguous. The option selected is not perfect. Some argument might be made, if business drivers and goals were reinterpreted or changed, that one of the other options might be better. It is closer to real life to have the option outputs overlap and, therefore, not point clearly to a single solution. Risk/uncertainty models are excellent vehicles to foster conversation and debate about input variables and to express the entire range of outcome possibilities, but it is very important to understand that such models rarely replace the decision-making process.

10 Risk Process to Identify Business Drivers, Maximize Value, and Determine the Value of Potential Expenditures

CONTENTS

INTRODUCTION

Conceptually, the unorthodox thread connecting all the concepts conveyed here is "linkage." This chapter is about linking actions to business goals and objectives, linking actions to a measure of value, linking actions to actual problems, and linking expenditures to reality.

Most good businesses generate and attempt to employ some sort of business plan. Corporate goals and objectives are parts of any rational business plan. To a widely varying degree, these goals and objectives are communicated to corporate personnel with the intention that the goals and objectives will inspire the workforce. Inspiration is intended to emanate from the establishment of hurdles to be overcome and targets to be achieved or exceeded.

Due to poorly constructed phraseology (too much "business speak"), inadequate communication skills and efforts, lack of line-of-sight for the worker between his or her job and the lofty and seemingly unrelated business goals, and other corporate-culture shortfalls, members of the workforce typically are at a loss to adjust their actions to achieve the espoused goals. Actions of workers are predicated on policies, political motivation, and past performance ("We have always done it like this") rather than on a focus on value.

In addition, when a serious misalignment of objectives and deeds becomes too obvious and painful to ignore, the typical reaction is to "fix" the things one knows how to fix, to the exclusion of enacting a remedy for the actual problem. This route is taken for several reasons. First, it is much easier and expedient to jump into action when fixing a problem that one knows how to fix — whether or not the fix actually yields any benefit. Second, there is a fundamental human tendency to wish to be ignorant of the actual problem. Links between actions and the real problems are weak or nonexistent.

Some corporations throw copious quantities of resources at a problem that are far in excess of the real or potential financial ramifications of the problem itself. Entire organizations have sprung up within a corporation to address real or perceived problems which could not possibly have been as expensive as the ineffectual solution decided upon. A link often seems to be missing between the financial threat from a problem and the price tag for a solution.

The point of including this chapter in the book is to offer a potential probabilistic solution to at least some of these common business foibles. In the following section, a very compact form of a scenario will be presented in which the main character must deduce a link between what action he should take and a narrow spectrum of corporate goals. In addition, the main character of the scenario will have to decide whether payment for a proposed solution to his problems is a financially reasonable action to take. A probabilistic model has been created as the centerpiece of the decision-making process.

THE PROBLEM

Bill is a production manager at Plastingcap, Inc. Management personnel at Plastingcap have recently become aware that a major airplane manufacturer is about to launch an effort to produce a new advanced-avionics aircraft targeted for sale to all major airlines. They have also learned that many older model planes will be updated to use the advanced avionics of the new aircraft.

Management believes that this news portends great things for Plastingcap because it has traditionally been the sole supplier of plastic knobs for avionics control panels. The advanced avionics systems would require a new line of plastic knobs that Plastingcap would be glad to supply. The vast majority of the avionics panels will be produced over the next 6 years.

Bill does not share management's unbridled zeal for the new project. Bill has, in the past, seen management embrace new projects simply for that reason — because they were new projects and were perceived as growth opportunities. However, almost half of the projects pursued did not, in the end, result in added value to the corporation. Bill believes that he should obtain from Steve, his manager, a clear elucidation of the corporate business objectives related to this project. Bill then plans to perform a simple economic evaluation of the new project to determine its degree of alignment with stated corporate goals.

At their meeting, Steve attempts to convey to Bill that he has a vision that includes corporate financial goals for the potential new project. However, the more Steve talks, the more convinced Bill becomes that Steve's financial goals are not clear in his own mind, and some goals are seemingly contradictory. In the meeting, Steve relates that it is important for the project to have a positive net present value (NPV) for the corporation and that the "Wall Street boys" would like to see Plastingcap have at least a 50% chance of realizing a profit in the second year of the project. Plus, top management would like there to exist a significant chance of realizing a cash-flow value of $1 million in the final year of the project.

Bill leaves the meeting feeling only a little better off than when he entered the meeting room. Bill knows that the new project will require at least some expenditure of capital funds specific to the modification of existing equipment and the purchase of new machines. Operating expenses also are expected to initially increase and rise slightly with time.

Because Steve's financial goals for the project were "clear as mud" to Bill and because Bill would like to generate some "scoping" economic estimates to form the basis of a second conversation with Steve, prior to doing an exhaustive economic analysis, Bill decides to put together a simple probabilistic model.

THE RISK/UNCERTAINTY MODEL

Bill decides to dust off a model that he has previously employed for simple economic analysis. Because the new project will not require the purchase of significant amounts of new equipment and because he wishes to keep the model as simple as possible for his upcoming discussions with Steve, Bill decides that he will, for now, ignore complicating factors such as depreciation. He deems the primary parameters of the model to be

First-year sales volume
Change in sales volume from year to year
First-year capital expenses
Change in CAPEX from year to year
First-year operating expenses
Change in OPEX from year to year
Price for the products
Discount rate
Corporate tax rate
Charges from vendors

The simple probabilistic model employed is

```
Revenue[0] = SalesVolume * Price;
CapExpenses[0] = CAPEX;
OpExpenses[0] = (OPEX[0] + (OPEX[0] – OPEX[0] * OPEXVar)) + Charge;
n=0;
while n < _NUM_PER do
    if n >0 then
        Revenue[n] = (Revenue[n-1] + (SalesVolume * SalesRate)) * Price;
        CapExpenses[n] = CapExpenses[n – 1] * CAPEXRate;
```

```
        OpExpenses[n] = (OPEX[n] + (OPEX[n] – OPEX[n] * OPEXVar)) +
          Charge;
      endif
      NOI[n] = Revenue[n] – OpExpenses[n];
      Taxes[n] = NOI[n] * TaxRate;
      CashFlow[n] = NOI[n] – CapExpenses[n] – Taxes[n];
      n = n + 1;
    enddo
    NPV = NPVsimple(CashFlow,DiscRate);
```

where

Revenue is the revenue, in dollars, generated by the project
SalesVolume is the sales volume of the project
Price is the price charged for the product
CapExpenses is the capital expenses for the project
CAPEX is the first-year capital expenses for the project
OpExpenses is the operating expenses for the project
OPEX is the first-year operating expenses for the project
OPEXVar is the uncertainty distribution associated with OPEX
Charge is the charge from a vendor
n is a counter used to indicate the current year of the project
_NUM_PER is the number of periods (e.g., years) in the analysis
SalesRate is the year-to-year change in sales volume
CAPEXRate is the year-to-year change in capital expenses
NOI is the net operating income for the project
Taxes is the amount of taxes, in dollars, paid
TaxRate is the corporate tax rate
CashFlow is the cash flow for the project, a time-series variable for a series
    of cash flows
NPV is the net present value for the project
NPVsimple is a system function that automatically calculates NPV from a
    series of cash flows and a discount rate
DiscRate is the discount rate

Armed with this set of initial parameters, Bill calls a meeting of his team to discuss the coefficients that should represent these variables in the risk model. The group addresses the parameters one by one.

POPULATING THE MODEL WITH DATA

Bill decided to begin the discussions with the known information. Plastingcap has been told by the aircraft manufacturer that the project will be 6 years in duration starting in the year 2000. Demand for the parts, if this project mimics others from the past, will increase through the sixth year. Bill's group knows that the discount rate is dictated to be 11% and that the corporate tax rate to be used in such calculations is 50%.

The group first addresses in earnest the projections of first-year sales volume. Experienced members of the group estimate that, given the number of new aircraft for which the manufacturer has firm orders, the number of parts supplied by Plastingcap will be between 10,000 and 15,000 in the first year. It is not clear, however, when the unrelated updating of older aircraft will begin. Plastingcap knows that the avionics manufacturer would like to begin as soon as possible. If retrofit of older aircraft is simultaneous with building the new aircraft, first-year demand for Plastingcap parts could reach as high as 30,000. A peakedness value of 5 is deemed appropriate to reflect their uncertainty regarding the most likely estimate for demand.

Next, the group discusses the question of increasing sales volume with time. Those on the team who have had previous experience launching other new products indicate that annual sales volume would increase over the 6-year project for several reasons. First, actual demand and use of the products typically increases through the final year. Also, subcontractors who use the products in the assembly of their equipment tend to stockpile extra parts for future use. Based on actual demand figures from other products, team members suggest that annual demand for the knobs would increase between 15% and 25% with a most likely increase of 20%. As with all variables in this scoping exercise, a peakedness of 5 is employed to indicate their middle-of-the-road confidence in their most likely estimate.

Based on the composition and configuration of the new plastic knobs, some maintenance procedures and updates to existing manufacturing equipment are deemed likely. Equipment engineers estimate that capital expenditure associated with the new knobs will range from a minimum of $200,000 to a maximum of $240,000 with a most likely expense of $220,000. They believe that they will incur at least some costs each year because of maintenance requirements and the inevitable changes made to knob design with time, but that, overall, costs will decline over time. Because final specifications for the knobs have not yet been forthcoming, they cannot have great confidence in their most likely estimate, so a peakedness of 5 is agreed upon.

The engineers responsible for generating capital cost estimates do so by utilizing a spreadsheet. Bill was able to lead the engineers into a discussion of a range for capital costs because they had to admit that they did not yet know the final specifications for the knobs and, thus, could not nail down a specific capital expenditure. However, their spreadsheet contains a single cell into which they input a single coefficient to represent their estimate of the increase in CAPEX from year to year. Because they cannot be persuaded to discuss a range for this parameter, Bill has to be satisfied with the deterministic estimate. They recommend the use of 30%.

Plant operations representatives who attend Bill's meeting espouse confidence in the fact that they can, with fair accuracy, predict what the operating expenses (OPEX) will be for each year for the new project. Their confidence stems from having had to predict, and then actually experience, the increase in OPEX associated with dozens of other projects in the past. They insist on providing a single OPEX value for each year and are not interested in putting a range on each deterministic value. However, Bill is able to get them to admit that there is some range around each of the single values for each year. They agree that the provided values might, in general, be high by as much as 20% or low by as much as 10%, but they consider the values to be precise. A peakedness of 5 is agreed upon.

Increases in OPEX from year to year for such projects have historically ranged from between 5% and 12% with a most likely increase of 7%. A peakedness of 5 is again selected due to the scoping nature of the current exercise.

Price was the final parameter to be addressed. The composition of the plastic used to manufacture the knobs has not yet been resolved. In addition, a variety of knob sizes will be produced. For these reasons and others, the group decides that it must come up with a reasonable range for price. Because the company likely will have to sign long-term contracts at a fixed price, its price estimate must be broad enough to account for the fact that the company will have to live with it for the next 6 years.

Economists and market-research representatives at the meeting estimate that the per-knob price will be a minimum of $1.50, a maximum of $2.50, with a most likely price of $2.00. Once again, a peakedness of 5 is used for this distribution.

A summary of the input data is shown in Tables 10.1 and 10.2. Figures 10.1 through 10.5 show frequency plots for those parameters that were represented by ranges of values and a peakedness value.

TABLE 10.1
Data Table for Risk Model

Variable	Minimum	Most Likely	Maximum	Peakedness
First-Year Sales Volume (#Units)	10000	15000	30000	5
Yearly Change Sales Volume (Dec. %)	0.15	0.2	0.25	5
First-Year Capital Expenses ($)	200,000	220,000	240,000	5
Yearly Change in CAPEX (Dec. %)		0.3		
Price for Product ($/Unit)	1.5	2	2.5	5
Discount Rate (%)		11		
Corporate Tax Rate (Dec. %)		0.5		
Vendor Charges ($)		0		
Variance in Yearly OPEX (Dec. %	0.8	1	1.1	5

TABLE 10.2
Time Series
Variable

Year	OPEX
2000	10000
2001	10500
2002	11000
2003	11500
2004	12000
2005	13000

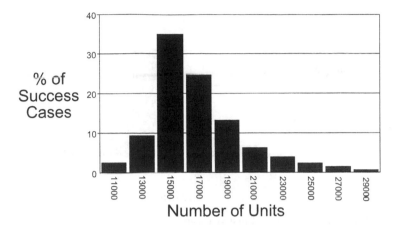

FIGURE 10.1   Frequency plot of first-year sales volume.

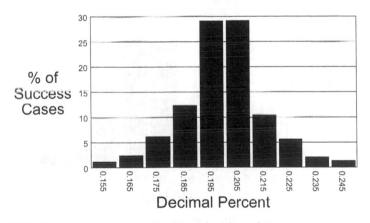

FIGURE 10.2   Frequency plot of yearly change in sales volume.

RESULTS FROM MODEL EXECUTION

Bill's risk and uncertainty model customarily produces four plots. Capital expenses, operating expenses, and cash flow time series plots are shown in Figures 10.6 through 10.8, respectively. A plot of net present value (NPV) is shown in Figure 10.9.

Figure 10.6 shows that, as expected, capital expenses will be relatively large in the first year. Capital expenses then drop off precipitously, approaching zero in the final year of the project. This, of course, is due to the requisite initial investment in new equipment and updating of existing equipment to get the project started.

Operating expenses shown in Figure 10.7 behave differently from capital expenditures. Because energy costs, wages, environmental-compliance costs, and other expenses are expected to increase with time, operating costs are also expected to increase with time.

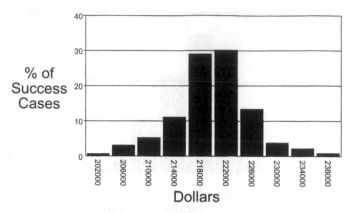

FIGURE 10.3   Frequency plot of first-year capital expenses (CAPEX).

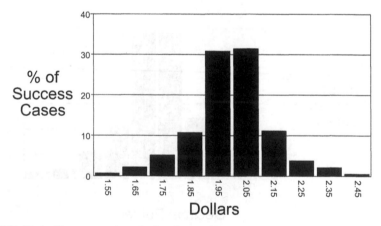

FIGURE 10.4   Frequency plot of price for product.

Figure 10.8 is a time-series plot of cash flow. As expected, high capital costs in the project's first year cause cash flow to be negative. However, Bill is certain that management is going to expect a positive cash flow in the second year of the project. Figure 10.10 shows the cash flow distribution for the second year. It is clear from this cumulative frequency plot that there is very little chance of realizing a positive cash flow in this year. In fact, the mean cash flow value does not rise above zero until the project's third year.

The cash flow bars in Figure 10.8 lengthen with passing years. This, again, reflects the increasing uncertainty regarding future cash flows. However, it should be noted that the bars are mainly lengthening on the "high" side. This, at least, is good news.

A plot of overall-project NPV is shown in Figure 10.9. As can be seen from the plot, there is a less than 5% chance that the project NPV will be negative. The project

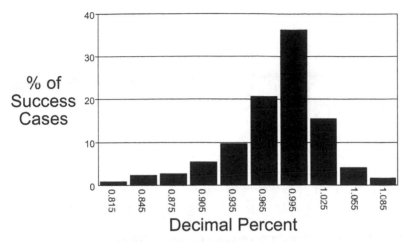

FIGURE 10.5 Frequency plot of yearly variance in operating expenses (OPEX).

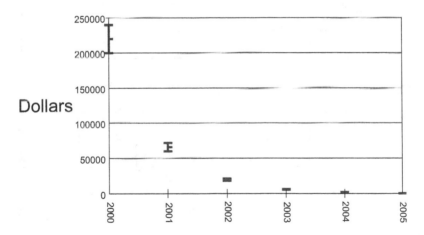

FIGURE 10.6 Time-series plot of capital expenses (CAPEX).

shows an "upside" NPV of relatively great magnitude, but the asymptotic nature of the cumulative frequency curve indicates that there is very little chance of realizing the high NPV values. The steep nature of the curve in the NPV range below $1 million indicates that the project will have a most likely value to the company that is less than $1 million. In fact, the net risk-weighted value for the project is near the half-million mark.

DETERMINING BUSINESS DRIVERS AND MAXIMIZING VALUE

Given the less-than-rosy cash flow plot and indications that NPV likely will be less than hoped for, Bill knows that he must take actions to attempt to improve the financial forecast for the project. Bill also knows from past experience that while a

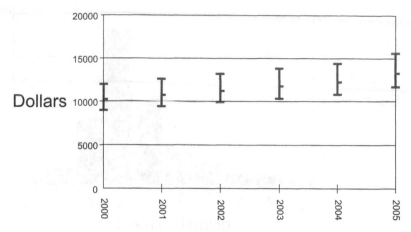

FIGURE 10.7　Time-series plot of operating expenses (OPEX).

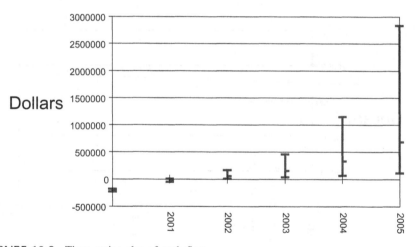

FIGURE 10.8　Time-series plot of cash flow.

given action on his part can enhance a particular financial measure, it might actually be detrimental to another. What he really needs is a much clearer expression from Steve of the company's financial strategy; however, Bill believes that this is not forthcoming.

Bill decides that he will generate action plans for three reasonable scenarios so that when he again meets with Steve, he can explain the various business scenarios and what might be done to accomplish each one. The first scenario will outline what business drivers need to be addressed in order to realize a positive most likely cash flow in the second year of the project. The second scenario will identify the business drivers that need to be acted upon if Plastingcap is to have at least a 50/50 chance of realizing a cash flow value of $1 million in the final year of the project. The final business goal for which Bill will identify crucial business drivers will be that of maximizing the NPV of the project.

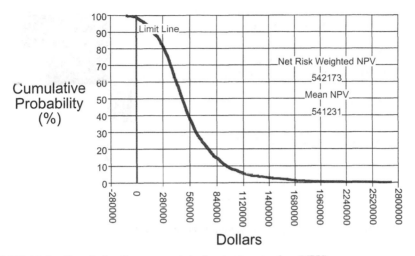

FIGURE 10.9 Cumulative frequency plot of net present value (NPV).

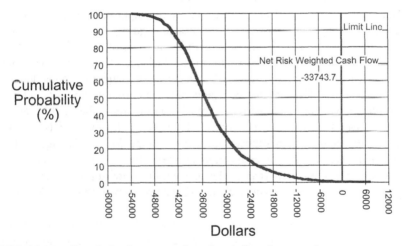

FIGURE 10.10 Cumulative frequency plot of cash flow for second year.

The software package Bill utilizes can perform sensitivity analyses (i.e., Spearman Rank Correlation analysis — see Chapter 18 for a discussion of this process) on model-output variables. For a time-series variable, the system allows a separate sensitivity analysis to be performed on any individual period in the time series.

Bill first executes a sensitivity analysis on the cash flow for the second year of the project. The resulting sensitivity plot is shown in Figure 10.11. From this plot, it can be seen that First-Year Sales Volume (FYSV) and Price for Product (PFP) are highly positively correlated with year-2001 cash flow as indicated by the length of the bars in the 0 to +1 correlation range (X axis of the plot). It also can be seen that FYSV accounts for more than 60% of the year-2001-cash-flow variance and that PFP accounts for more than 30% of the variance. This tells Bill that if he wants to

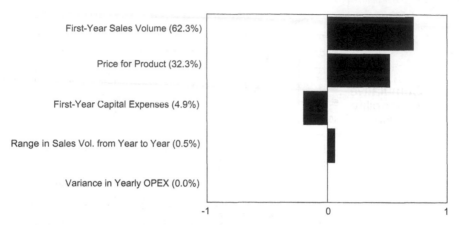

First-Year Sales Volume (62.3%)

Price for Product (32.3%)

First-Year Capital Expenses (4.9%)

Range in Sales Vol. from Year to Year (0.5%)

Variance in Yearly OPEX (0.0%)

-1   0   1

FIGURE 10.11   Sensitivity plot for second year of project.

improve the cash flow for the second year of the project, he must primarily focus on steps that will boost first-year sales volume.

Note that in Tables 10.1 and 10.2 there are ten input variables, but only five of those variables appear on the sensitivity plot in Figure 10.11. Inspection of Table 10.1 and the mathematical nature of sensitivity analysis of this type should make it clear that only variables that contribute to the variance of the output variable (i.e., only input variables that are expressed as a range of values) are considered in the sensitivity analysis. The fact that constants such as corporate tax rate and discount rate do not show up on the sensitivity plot does not really bother Bill because these are things over which he has no control. However, Bill now regrets not pushing harder on the engineers to supply him with a range for the Change in CAPEX variable. Bill also has constructed his model code so that OPEX, which is represented as invariant values for each year of the project, does not appear on the sensitivity plot. Bill makes a note to consider changing the program so that OPEX appears on the plot.

To increase first-year sales, Bill will propose to Steve that they might take one or all of three actions. Plastingcap might make every effort to keep the price for the knobs at the lowest possible level in order to increase sales. A second ploy might be to increase expenditures on advertising, but this likely will have little effect because the knobs are not intended for sale in the greater marketplace. The third suggestion Bill will make to Steve is to hire a firm that can increase Plastingcap's sales to foreign airlines and airplane manufacturers. Bill knows of such a company and will immediately investigate this option.

To determine what steps would be necessary to realize a reasonable chance of attaining a last-year cash flow of $1 million, Bill generates a sensitivity plot for the final year of the project. This plot is shown in Figure 10.12.

As this plot shows, if Bill wants to positively affect the cash flow in the final year of the project, his priorities, relative to those for the first year of the project, must change. Again, it can be seen that PFP and FYSV are positively correlated

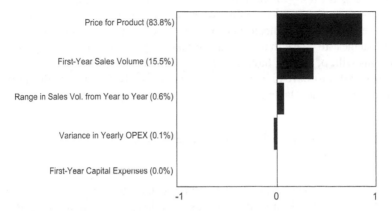

FIGURE 10.12   Sensitivity plot for final year of project.

with year-2005 cash flow. However, PFP is by far the variable that most impacts year-2005 cash flow, accounting for more than 80% of the output-variable's variance. Although Bill can identify the primary business driver, this reversal of drivers (for the second and last years of the project) poses a significant problem.

To increase second-year sales, one of the options Bill was going to suggest to Steve was the ploy to keep prices as low as possible. Bill now realizes that if he pursues the low-price option, he will actually significantly damage his chances of realizing a $1 million cash flow in the project's final year. So, it seems as though he can accomplish one business goal or the other, but probably not both. Furthermore, accomplishment of one goal might actually have a significant negative effect on the other. Bill knows that he must make this clear to Steve in their upcoming meeting.

Bill finally produces a sensitivity analysis for NPV which appears in Figure 10.13. Bill is greatly relieved to see that the top two significant parameters for NPV

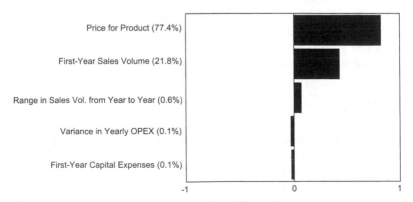

FIGURE 10.13   Sensitivity plot for net present value (NPV).

are the same as for last-year cash flow and, respectively, account for about the same amount of variance as in the cash flow situation. Given that sensitivity plots for NPV and last-year cash flow indicate that product price is the principle business lever, Bill will suggest to Steve that Plastingcap take steps to maximize the price for the knobs. This will, of course, have a detrimental effect on second-year cash flow, but Plastingcap will just have to convince the financial analysts that negative cash flows in early years are acceptable given the long-term strategy of the company.

DETERMINING THE VALUE OF POTENTIAL EXPENDITURES

Although FYSV is second in effect to PFP on the last-year-cash-flow and NPV sensitivity plots, it is still a significant business driver. Bill, therefore, pursues his idea to enlist the services of a firm that could help Plastingcap increase sales overseas.

Bill meets with Ron, who is a representative from a well-known consulting firm that can provide the foreign-country services Bill requires. After several meetings, Ron presents Bill with a plan that could double Plastingcap's sale of knobs by significantly increasing foreign sales. Ron indicates that he could double first-year sales and maintain the doubled sales throughout the life of the project for a fee of $200,000 per year. At first, the $200,000/year figure to double sales sounds reasonable to Bill. However, before agreeing to the proposal, Bill decides to "run the numbers" in his risk model.

The FYSV figures shown in Table 10.1 indicate that first-year sales are expected to range from a minimum of 10,000 to a maximum of 30,000 with a most likely value of 15,000. The net NPV to the company for such sales volumes is between $500,000 and $600,000, as indicated in Figure 10.9.

Bill reruns the model using FYSV figures of 20,000 for a minimum, 30,000 for a most likely, and 60,000 for a maximum to reflect the doubling of sales volume. He also adds the $200,000 yearly charge to each of the yearly OPEX values. The resulting NPV plot is shown in Figure 10.14.

The net-risk-weighted NPV for the doubled sales volume is shy of $1,000,000. There is about a 5% chance that the NPV will be negative. In addition, the new cash-flow plot for the final year of the project (not shown) indicates that there is just over a 50% chance of achieving a last-year-cash-flow value of $1,000,000. All of this looks pretty good to Bill, but he will present and discuss the results with Steve before making any decisions.

SO, SO WHAT?

There are quite a few lessons to be taken from this chapter. First, it is not uncommon for people like Bill to be charged with "successfully" executing a project without the existence of a clear definition of success. In Bill's case, he could be deemed to be successful if he turned a profit in the second year, even though this would reduce the chance of $1 million cash flow in the final year and would decrease the NPV.

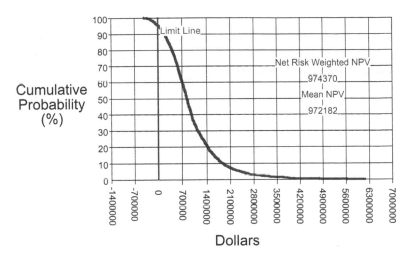

FIGURE 10.14 Cumulative frequency plot of net present value (NPV) for doubled sales volume.

Conversely, Bill might successfully maximize final-year cash flow and NPV, but those outcomes might have a detrimental effect on second-year cash flow.

In such a scenario, it would be absolutely essential for upper management to clearly understand that to pursue a maximum NPV might depress early earnings. If this concept is not clearly conveyed to management, the pursuance of a successful strategy (maximizing NPV) could cause the project to be canceled early because of poor early year performance.

Sometimes those "in charge," such as Steve, do not have a good grip on just what the financial goals of the corporation are. Therefore, people in Steve's position sometimes are not capable of making "ground-level" decisions regarding specific strategies for specific projects. It often is up to people in Bill's position to scope out the various approaches and to clearly communicate approaches and associated ramifications.

It should be evident from the example presented in this chapter that various risk/uncertainty related tools can be employed to sleuth out the major business drivers for a project. Often, as in this case, several sets of business drivers need to be identified — each set relating to a separate business strategy.

Another connection that should be made is that between business strategy and passing time. The example presented here was so constructed that the primary business drivers which would benefit the project in early years were different from those (or a different order, at least) which would most significantly impact the project in later years. When considering time-dependent financial measures such as NPV (created from a series of cash flows), it is essential to define the strategy of a business in terms of time. To simply state, for example, that "we want to maximize cash flow" is not sufficient.

Finally, it was made clear that risk/uncertainty-related techniques can be used to determine the value of potential expenditures, such as the charge for services to increase sales volume. While it is true that such calculations can be performed without a stochastic risk model in a deterministic manner, the benefit of being able to view the entire spectrum of possible outcomes and related probabilities makes using such a model worthwhile. Decision making is always enhanced when the entire range of possibilities is considered.

11 Summary

CONTENTS

OTHER APPLICATIONS

In preparing this book, I suffered from a wealth of options. Over the years, my colleagues and I have written or helped others construct literally hundreds of risk and uncertainty models and processes. I cannot think of a major aspect of business that we have not addressed.

The risk and uncertainty applications that constitute the subject matter of the previous ten chapters were culled from a plethora of possibilities. In selecting the applications delineated in those ten chapters, I endeavored to select a set of examples that would address a broad spectrum of business situations. I also intended the selected examples to demonstrate to the reader that risk and uncertainty techniques can be applied to the problems with which the reader has experience. I hope readers can "see themselves" in one or more of the chapters.

Because of space limitations and other practical considerations (such as the price of this book), it was necessary to limit the number of examples and chapters to about ten. This limitation, of course, necessarily precludes many very interesting and useful examples from being incorporated into this volume. I found this to be exceedingly annoying and decided that if I could not appropriately expand these excluded examples in a chapter, I would at least mention some of these applications in this summary chapter. If readers decide they would like to discuss or obtain examples of applications listed below, they should feel free to contact me and I will do what I can.

INSURANCE EXAMPLE

Typically, people in the insurance business attempt to decide how to establish reasonable premiums while considering rare but expensive events. Such individuals or groups usually need to determine the effect of highly unlikely but disastrous events on premiums.

For example, folks in the pipeline-insurance business have to consider the impact of the rarely occurring but very real problem of pipeline rupture and the related loss of property and life, fire, environmental disaster, and other major calamities. Likewise, firms that insure houses in certain parts of the country have to consider the possibility of damage from hurricanes or tornadoes. Many relatively sophisticated risk and uncertainty models have been constructed to aid in determining the effect on premiums of such potential mayhem and other more mundane considerations.

On the other side of the insurance fence, those who might purchase insurance typically have to decide whether it is cost effective to insure against various events. Is it, for example, cost effective for a plant manager to carry insurance against the improbable situation in which his plant would be flooded (with water)? If flooded, the plant would shut down and suffer loss of revenue. Is it prudent to simply suffer the consequences of shut down, or to insure against revenue loss and cleanup? Many a risk model has been constructed to help make such decisions regarding various areas of business.

Pricing Example

Growth and generation of profits are fundamental precepts of most businesses. To grow, it is often necessary for a business to expand into new product lines. Profits from sales of new products depend, in part, on the price that can be established.

Prices can be cyclic. Depending on market conditions, competition, demand, inventory, and other factors, prices can over time swing wildly through a broad range. Such fluctuations and their duration can make it difficult to predict a potential new product line. Risk and uncertainty models can be designed and constructed to address the problem of determining the value of a product for which the price is subject to cyclic variation.

A model of this type might be composed of variables such as

- The number of units sold in peak-price times
- The number of units sold in trough-price times
- The price/unit in peak-price times
- The price/unit in trough-price times
- CAPEX in peak-price times
- CAPEX in trough-price times
- OPEX in peak-price times
- OPEX in trough-price times

In addition, such models have been built to include parameters to control the switching from peak-to-trough times (or trough-to-peak); the duration of the cycles; chances of failure associated with political, environmental, economic, and technical events; and a host of other considerations. Time series plots and plots of financial indicators such as NPV that result from such models can be used to predict the frequency, timing, and severity of price swings and the economic impact on the corporation. Because all of the parameters are variable, such models facilitate simulation of many business scenarios. Plots of "summary" financial measures such as NPV and associated risk-weighted values allow quick and easy comparison and ranking of scenario consequences.

ENVIRONMENTAL APPLICATIONS

When environmental risk and uncertainty models and processes are considered by corporate practitioners, the aim of such applications typically is to assess and define mitigation steps and costs associated with an environmental infraction of some sort. Higher-than-allowed exposure rates, toxic spills, and airborne contaminants are representative problems. To feed the design and construction of many such models, data must be collected for variables such as diffusion rates, porosity, toxicity, demographics, air flow, exposure frequency and duration, and other measurable parameters. It is not uncommon for environmentally oriented models to focus on such physical and measurable variables. However, environmental considerations need not be constrained to physical or economic concerns.

The design and construction of "broad spectrum" risk and uncertainty processes collect and integrate information on what might be considered unconventional parameters. The goals of such models can be to influence the assessment and to potentially mitigate actions by integrating the impact of social, political, legal, infrastructure, and other concerns. For example, a corporation may be considering doing business in, what is to them, a new country. Prior to plunging ahead, an assessment of environmental conditions of a broad spectrum might be prudent. The company might consider whether there is a clear permit-granting process for sanctioning the legal documents required. Availability of medical facilities and staff to treat corporate personnel might be assessed. The corporation's environmental image should be considered, as well as the activities and aims of environmentally based opposition groups. Access to land and transportation infrastructure in general could be considered, as well as reputational issues and political ramifications with other governments. The list of parameters is, of course, virtually endless.

Endless though it may seem, the best, most comprehensive, and useful environmental models and processes have incorporated a heavy dose of such parameter sets. Data gathered for such variables are represented in models as distributions, chances of failure, dependence parameters (see Fundamentals of Risk Assessment section of this book), and other influences and are used to impact financial calculations and mitigative actions. Qualitative as well as quantitative influences can be considered and integrated into such models. Coefficients representing model or process output parameters, having considered such variables, are exceedingly useful in the decision-making process and better predict real-world conditions.

LEGAL APPLICATIONS

Chapter 1 of this book makes an example of a relatively simple legal application that considers various ways that a decision tree might be converted to a probabilistic model. A class-action suit served as the focus of the example. As any attorney will attest, real-world legal decision making can be much more trying (pun intended).

Everyone in the legal game is aware of the usual "cast of characters" that constitute, for example, a litigation model. Variables such as the probability of a favorable ruling in a given jurisdiction, court costs, attorney fees, cost of judgments, punitive damages, probability of a mistrial, and other more mundane considerations are typically employed to calculate a range of (financial) impact and associated probabilities.

A template of parameters considered for such models serves as a basis for discussions with clients, promotes common understanding of input data and output consequences, and affords a quantitative basis upon which to make a decision. Corporate attorneys and others today, however, have to address a much broader spectrum of considerations in the decision-making process. For example, company A might be involved in litigation with company B, a supplier of raw materials for company A. Although a boilerplate quantitative risk and uncertainty model employed by company A might indicate that there exists a good probability that company A will win a significant judgment, attorneys for company A still might not be eager to drub company B in court.

Astute attorneys will, in their risk and uncertainty model, consider a wide variety of business-related concerns. For example, if company A wins the case against its supplier, how will that impact company A's ability to obtain raw materials? This has to be considered not only for company B, but for all raw-materials suppliers who are watching attentively to see how they can be expected to be treated by company A. If the suppliers like what they see, company A might find itself a "legal winner" but a business failure.

On the other hand, if company A fears that it will lose suppliers and, therefore, business by winning, it might consider some sort of settlement with company B. Attorneys realize, however, that such a settlement, unless deftly handled, could lead to a host of "copycat" claims by other suppliers.

Reputational aspects also must be considered. Company A might be the proverbial "500-pound gorilla" in the industry dwarfing company B. The public reputation of company A, and therefore sales, might be irreversibly and negatively impacted by a well-publicized beheading in court of company B. Given the reputation consequences of a victory, attorneys have to consider in their model whether they can afford to win.

The aforementioned nonconventional considerations and others should form the backbone of any modern risk and uncertainty legal model. It often is the case that the impact of consideration of such parameters outweighs the financial consequences of integration of more conventional variables. When business, political, reputational, and other data are considered in a comprehensive analysis, decisions resulting from that analysis can be better aligned with corporate goals and strategies.

SECURITY APPLICATIONS

For corporations, governments, and individuals, security concerns are paramount. In today's world, it is not enough to simply consider the probability and range of possible consequences of breaches of security. Because the scope of most major businesses is global, contemporary security considerations are a far cry from what they were when the typical business needed to consider only domestic issues.

In the past, corporations could compile and distribute comprehensive corporate guidelines for security. Most of the rules and regulations put forth in such documents were either commonsense items or procedures that were legally prescribed. In today's global business environment, issuance of such "one size fits all" security proclamations is woefully insufficient. Adoption of specific security measures in one country, for example, might violate religious, cultural, or legal precepts in another. Enactment of specific security procedures might be interpreted as insulting to a host government, or could be construed as an act of favoritism by a neighboring country.

Well-designed and constructed comprehensive risk and uncertainty models that address security concerns must necessarily embrace considerations such as those outlined above. To accomplish this, two approaches are common. Some corporations lean toward the design, construction, and employment of a large and comprehensive model that can be applied to any country and security situation. Such models have the advantage of generating a fixed set of output parameters that can be used to compare and rank results from individual assessments. In addition, this type of model allows corporations to keep all of their security issues together, avoiding the fragmented nature of multiple models. The "dark side" to such models is the fact that they become unwieldy for the user (they may contain many parameters that do not necessarily apply to a given situation) and are notoriously difficult to maintain due to the comprehensive and convoluted nature of the relationships between the considered parameters.

An alternate approach is to generate a unique risk and uncertainty model for each country or for a specific group of countries. This approach tends to keep the models relatively compact, nimble, easy to use, and focused; however, a number of models will have to be built and maintained. In addition, unless the models are coordinated, it is exceedingly likely that the various models will produce output plots and data that cannot be compared.

In spite of implementation foibles associated with such models, the fact remains that risk and uncertainty processes and models that address security issues need to be increasingly sophisticated in order to produce probabilistic output that is relevant to today's global business environment. It is more of an art than a science to build a security assessment model that is simple to distribute and use and yet sufficiently comprehensive and sophisticated to address the complex security issues of today. Such models can and have been constructed.

IT IS MOSTLY THE PROCESS — NOT THE TECHNOLOGY

Over the years many methods for dealing with risk and uncertainty analysis have been investigated — decision trees (all sorts of variations on the theme), linear programming, genetic algorithms, and other less well-known approaches (see Selected Reading). Many of these alternatives to simulation techniques (mainly, Monte Carlo) have much more appeal than the Monte Carlo method, which has been around in a practical form, since the end of the second world war. I approach most, but not all, risk and uncertainty modeling with the Monte Carlo method, not because it is technically superior (it is not), but because it best fits the business processes in which it has to be embedded.

There existed, in the past, two primary drawbacks to the Monte Carlo approach. First, it is iterative. Say what you like about Latin Hypercube and other speed-enhancing techniques, Monte Carlo analysis on vintage computers was just flat-out mind-numbingly time consuming when even a moderate number of iterations (say, 1000) were performed. Today, even relatively inexpensive desktop processors are of such efficiency as to make moot the run times for most efficiently constructed Monte Carlo models. This is especially so when the models are "stand alone" (the type I prefer to use) rather than those that "drive" spreadsheets.

The second technical problem that vexed Monte Carlo analysis was that of constructing distributions. When I first began to roll out fledgling Monte Carlo models, it became glaringly obvious that people could tell me intrinsically what a parameter distribution might look like (they had a mental model). However, they could not manipulate means, standard deviations, kurtosis values, alpha-1 and alpha-2 parameters, nu values, etc., to actually build a distribution shape to match their mental image. Most Monte Carlo-based software packages that require such manipulation make the assumption that, for example, an engineer with knowledge of such subject matter is the target user. In my experience, rarely was that the case. So, I set about developing algorithms that would require from the user no statistical input and would yet allow them to construct a distribution of virtually any shape (to fit their mental model). The general philosophy regarding this approach is discussed in the second half of this book.

With these two hurdles overcome (run time and distribution building), there remained no major technical impediments to the practical use of Monte Carlo analysis in most aspects of business. However, some significant human-nature and practical problems remained.

The problems relating to design and implementation of risk and uncertainty processes and models in the real world are presented in the first book and are not repeated here. However, political, organizational, cultural, and human-nature impediments are not easily exorcised. For example, it is exceedingly difficult for data providers to realize and express the real range of values associated with a model parameter. Special facilitation techniques have to be applied to even begin to extricate from people just how "big" or "small" or "good" or "bad" an entity or situation might be. Getting people to realize and express dependencies between variables, apply chance of failure, or express a consensus definition of the problem to be addressed are all vexing problems that exist regardless of the technology applied to process the data.

I reiterate here that I favor Monte Carlo analysis (now that the two major hurdles have been overcome) because I find it best fits the business to which it has to be applied. However, it is not the technology as much as the implementation process that has grown up around the technology that is so helpful.

Over the years, I have run into the same guy time and again — he is always in a different body, but he is the same guy. This is the guy, typically a manager of some variety, who is not in favor of "this uncertainty stuff" and makes it abundantly clear to me that such processes are a destruction of value because he would have made the same decision without "all this hocus pocus." When this opinion is not-so-delicately expressed, I always respond that he likely is right — that the same decision might have been made without a probabilistic analysis. However, I also attempt to point out some of the major advantages of employing the stochastic approach.

First, the prescribed approach promotes consistency. That is, although the manager might arrive at a decision on one day, a similar decision on a subsequent day might or might not consider all of the necessary and critical elements. Using a model to force consideration of the salient elements attempts to ensure that the possibility of overlooking a critical item is minimized. Consistency leads to fewer mistakes and a better educated workforce.

Another advantage is that the process of combining the various parameters can be explained. When people understand how a decision was made, "buy-in" is enhanced from those impacted by the decision. An analysis of a situation made in the head of a manager rarely is documented or completely explained so that others can fully comprehend.

Learning with time and experience is yet another benefit. A well-conceived and implemented probabilistic process should archive both input and output data. Such archived data can be a significant aid in generating reasonable input data during future executions of the model.

The most significant benefit of the stochastic approach, and especially of the Monte Carlo technique, is the conversation that it fosters. When attempting to populate a model with the requisite data, a skilled facilitator can lead individuals or groups through discussions of the parameters. Such conversations often are more illuminating than the ultimate output from the model. In populating a Monte Carlo-based model with data, discussions regarding the real-world ranges of coefficients that represent parameters and the interplay of those parameters is most enlightening. At the culmination of such discussions, all parties should have a firm understanding of the model input data.

ACCOMPLISHMENT OF VISION GENERATES REAL RETURNS

Of course, the "bottom line" on this risk and uncertainty stuff is how it positively impacts the "bottom line." Real benefits — financial and other — can be realized when risk and uncertainty technology is combined with a well thought-out and articulated vision. Two actual examples are offered here.

EXPLORATION EXAMPLE

In the oil and gas exploration game, a prospect (usually a prospective well to drill) typically sports an economic chance of success in the 15% to 30% range (I am speaking here of true exploration wells, not of development or infill-drilling wells). That is, typical wildcat exploration wells can have less than a 30% chance of making money. In a particular company, it was reasoned that if they could uniformly assess their yearly portfolio of more than 100 exploration wells, they might be able to rank and prioritize the opportunities. Comparison of the prospective wells would be made on the basis of a host of risk-weighted values resulting from the risk analysis.

The company accomplished the task of assuring consistency and realistic input to the model by enacting the "police" methodology (see Chapter 3). That is, before any prospective well could be submitted to the exploration portfolio for potential funding, it was required that a board of experts review the model input data.

Positive results from implementation of this process were numerous. First, whether a prospective well was funded or not now depended much less on "salesmanship," corporate nepotism, and politics. Second, all opportunities across the world were compared on a consistent basis, allowing comparison and ranking. When the cold eye of assessment was cast upon the portfolio of opportunities, it was found that only a fraction of the prospects that would have previously been considered for funding were actually quality opportunities. Several years of assessment and archiving of data helped them characterize quality prospects.

Culling the prospects resulted not only in drilling fewer wells, but also in drilling the very best ones. After a few years of practice, the company was drilling about one quarter fewer wells than it had prior to implementation of the process and was realizing more resources from a smaller set of high-quality wells. In addition, the stochastic process facilitated, at the beginning of a fiscal year, the accurate prediction (eventually, to within about plus or minus 10%) of the level of resources that would be realized from the annual portfolios of opportunities. More resources, of course, meant more revenue. Fewer wells translated to huge savings in costs. Accurate prediction translated to more efficient distribution of resources and spending of capital. Combined savings and increased revenues were measured in the hundreds of millions of dollars.

Maintenance/Construction Example

A manager at a chemical plant had a vision. The vision was driven by shrinking capital budgets and, seemingly, increased inefficiency in construction and maintenance projects. There were several problems.

First, personnel proposing projects were competing for a slice of a capital-budget pie that was ever shrinking. Therefore, to make a project appear attractive, it had to appear cheap. However, when the project was actually pursued, a huge "contingency" budget was allocated because initial estimates of costs were ridiculously optimistic. Second, some capital spending was wasted on projects that, after invocation, were canceled. This, of course, was not efficient use of capital.

The manager's vision was to obtain more accurate estimates of project costs. In addition, he envisioned comparing and ranking projects not only on cost, but on strategic alignment. Inclusion of strategic considerations, he mused, might cut back on projects that were initiated only to find that they did not align with local or corporate goals.

The manager called in the risk and uncertainty experts to work with select members of his staff. The team was to design a probabilistic model and process that would translate the vision into reality.

A long (about 6 months) and arduous process began, which culminated in the generation of a comprehensive risk/uncertainty model and model-implementation process. The model itself considered both economic parameters and measures of strategic alignment. A "police" system was established. In this case, multiple review boards were set up because it was deemed that, for example, a maintenance project was sufficiently divergent from a new-construction project to warrant a distinct cadre of reviewers. Projects could not be funded unless first reviewed.

"Policing" the model inputs solved the problem of underestimating costs. This helped the manager get more realistic estimates and, therefore, a better grip on how many projects he could handle. In turn, contingency budgets shrank considerably. Because he now had a better understanding regarding the number of projects likely to be pursued, great savings related to company and contractor personnel were realized.

The measures of strategic alignment included in the model reduced to almost nil the number of projects begun and then canceled due to misalignment with local or corporate goals. A common financial output parameter from the model allowed comparison and ranking of opportunities. Combined savings were measured in the millions of dollars.

Section 2

Fundamentals of Risk Assessment

12 Building a Consensus Model

CONTENTS

WHAT IS THE QUESTION? — MOST OF THE TIME AND EFFORT

It happens almost every time. Someone from a group in the organization will call me or e-mail to request help with a risk assessment model or process. When I arrive on the scene, typically there is a conference room full of people to greet me. Folks from all areas are gathered — managers, political specialists, economists, engineers, other scientists, environmental specialists, legal counsel, and others. I liken this gathering to the three blindfolded persons and the elephant. If you were to ask these three persons: "Do you have an elephant?" their universal answer would be "Yes, we certainly do." However, if you asked each person to describe the beast, the one holding the tail would say it was like a rope, the one touching the leg would indicate that it was like a tree trunk, and the one holding the ear would relate that it was like a big flat thing.

Initial meetings with groups attempting to build a consensus risk assessment share some characteristics with the three blindfolded individuals. If you ask: "Do we need a risk assessment?" the universal answer will likely be "Yes, we do." However, if you ask: "Who thinks he or she can state in one sentence (two at the most) just why we are here today?" many hands in the room will go up. Each person has his or her idea of just what question is to be answered. I can tell you from experience that there will be little commonality to their answers. An important function of the facilitator is to attempt to meld the diversity of opinions and viewpoints into a risk model — a consensus risk model — that everyone agrees is sufficiently comprehensive, rigorous, and flexible to accomplish the task at hand. This is a tall order.

221

The very first thing to do when facilitating such a group is to ask someone (who is not the boss) to write on a "flip chart" or board, in a single sentence, just what he or believes the problem to be. When this has been done, ask the rest of the group if they agree with exactly what has been written. If they have no other opinion (which they will not if the boss does the first writing), you will be in the presence of a very unusual group. Typically, engineers see the question as having to do with the engineering aspects of the problem. Economists likewise see it as fundamentally an economic argument. And so it goes with each person and each discipline. Arriving at a consensus on exactly what problem (or problems) needs to be solved is a tricky proposition.

CONSENSUS MODEL

When multiple projects of similar type must be evaluated (like the previously discussed plant-construction projects), a consensus risk model should be constructed with the aim of applying this model consistently to all projects of that type.

The ability to rank, compare, and portfolio manage risk-assessed opportunities is predicated on the assumption that the assessed opportunities can, in fact, be compared. Comparison of opportunities requires that all opportunities be compared on the basis of one or more common measures. Typically, these measures are financial, such as net present value or internal rate of return. However, measures that are not financial also are common. Measures such as probability of success (technical success, economic success, etc.), product production, and political advantage can be used to compare opportunities.

As an example, let us consider the situation in which a business manager is charged with the task of deciding which of several competing product lines should be selected for marketing. The product lines may be diverse. One line may be footwear, one hunting gear, one basketball equipment, and so on. The manager may decide to compare these opportunities on the basis of net present value to the company. The challenge, however, is to decide whether to allow the departments for the product lines to construct their own assessment models or to attempt to generate a consensus model.

If the comparison of competing product lines is likely to be a unique or rare event, then the manager might decide to allow each department to devise its own assessment model as long as the result of the analysis is the net present value of that product line. If this is the path taken, it will then be up to the manager to ensure that all critical elements (taxes, labor costs, other expenses, etc.) are considered in each department model. Such an undertaking can be a daunting task. This method of comparison may be most efficient in the case of assessing events that are not likely to occur again.

The manager, however, may decide that he or she will be confronted many times with making such decisions. In this case, it is more efficient to attempt to build a consensus model which can be applied to the three product lines presently being considered and to all other product lines that may be considered in the future. A graphical example of these two contending philosophies is shown in Figure 12.1.

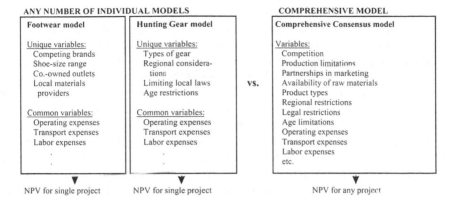

FIGURE 12.1   Examples of plans for multiple-individual models vs. a consensus model.

To build a risk model capable of considering all the intricacies of a multitude of product lines, the manager will have to assemble critical representatives from each department and engage them in the risk-model-building process. This can be a time-consuming and sometimes a nearly politically intractable task. However, the effort will pay off with a risk model that can be applied repeatedly to new opportunities or sets of opportunities. Those opportunities can then be compared with options already assessed. Portfolio management is the result.

It should be expected that the risk model will be an evolving entity. With the repeated application of the model to new and slightly different situations, the model will grow and evolve into a more robust and comprehensive assessment tool. Appropriate resources for the maintenance and enhancement of such a model should be considered essential from the beginning of this process.

GROUP DYNAMICS

The real trick in arriving at the consensus solution is to do it without hurting anyone's feelings or having someone feel as though his or her contributions are not being justly considered. Many approaches are used to resolve this type of situation; however, they all share the common thread of orchestrating group dynamics.

The first piece of advice I generally give to a group attempting to reach consensus on a question or model is to use a facilitator (discussion/model-building leader) who does not have a stake in the outcome of the project. The best facilitators are those individuals who

- Have good "stage presence"
- Have good interpersonal skills
- Understand the process of orchestrating group dynamics in the risk-model-building arena
- Completely understand risk technologies

In fact, the best facilitators are those who exhibit the attributes listed above and who have no specific knowledge concerning the problem to be solved. Relative ignorance not only imparts an aura of impartiality, but allows the facilitator to ask fundamental questions that otherwise may be ignored or the answers to which would be taken for granted. Discussion of fundamentals many times can be of great importance in the process of reaching consensus.

For example, when there is a gathering of individuals concerning a project for which a risk model is to be generated, many of the individuals know one another. This can be both beneficial and detrimental. The benefit is derived from the feeling of comfort that comes with familiarity. This comfort factor, shared by the people in the room, can be used to move things along because introductions are not necessary and "meeting-new-people jitters" is not a factor. Contributions to the discussion are more freely given in familiar surroundings. However, familiarity also can work against the situation.

In many cases, people who know one another "carry baggage" with regard to past relationships. In addition, there may be competing entities in the room. This may take the form of competing individuals ("only one engineer is going to get to work on this"), competing groups, or other forms of competition. This and other factors such as downright "bad blood" have to be considered when attempting to facilitate a risk-model-building exercise and must be mitigated in a constructive manner. There are many group dynamic processes that can help.

WRITE IT DOWN

After the ice has been broken by the first person expressing his or her written opinion as to just what is the problem, the facilitator should ask those who take issue with the written statement to write on a piece of paper, in a single sentence, just what they think is the problem. If the facilitator allows verbal expression of the problem(s), chaos results because, invariably, an argument or heated discussion ensues. The facilitation steps to follow next are

- Have volunteers, one at a time, come to the board and write down their expressions of the problem
- Following each statement, direct the group to find some common elements in previous statements and the newest one
- Keep a separate written list of the common elements in full view

Figure 12.2 depicts a table that might result from the process described above. In the example shown in Figure 12.2, we are really lucky that a common thread concerning the economics of the project could be found in all points of view. To calculate the net present value (NPV) of the project, we will of course consider the engineering, environmental, economic, political, and other parameters, but at least we now have a common and agreed-upon goal for the risk model.

If a reasonable list of common elements has been compiled, then the next step is to engage the group in attempting to generate a concise (i.e., short) statement that captures the salient common points. This process does not have to result in a single statement. More than one problem may be recognized; however, bear in mind that

---

PLANT-EXPANSION MEETING - JUST WHAT IS THE QUESTION?

Individual Statements of Problem

#1 - Engineer 1 - What is the best affordable construction method for the plant-expansion scenario selected?
#2 - Engineer 2 - What is the plant-expansion scenario that yields the highest net present value?
#3 - Economist - Regardless of the plant-expansion scenario (i.e. technical plant-expansion details), what is the net present value of expansion of the product line?
#4 - Environmental Engineer - What effect will spills and other environmental concerns have on the profitability of the project?
#5 - Political Consultant - How will the timing and economics of the project be affected by governmental attitudes and requirements?

Dynamically-Kept List of Common Points

Common points between statements 1-2: Scenario and affordable plant expansion
Common points between statements 1-2-3: NPV and expansion
Common points between statements 1-2-3-4: Profitability (i.e., NPV)
Common points between statements 1-2-3-4-5: Economics (i.e., NPV)

FIGURE 12.2 Compilation of ideas for the purpose of determining just what question the risk assessment will attempt to answer.

separate problems may require separate risk assessments. It is the facilitator's job to ensure that all contributing parties in such a process are recognized and that all ideas are given a fair hearing. In this case, we may settle on the statement

What is the cash-flow situation and NPV of a selected plant expansion scenario?

If a high-level and general risk model is constructed, then several, if not all, plant-development scenarios will be able to be evaluated with the same model. This is now the challenge.

SORT IT OUT

If no, or very few, common elements are identified (the case when people have very different ideas as to just what the problem is), then a sorting mechanism must be used to arrive at the most important of the issues expressed. This sorting can take many forms. A hierarchy diagram (see Selected Readings for this chapter) is a commonly used tool. In a typical hierarchy diagram, the policies are listed at the top, the strategies in the middle, and the tactics at the bottom. The typical form for a hierarchy diagram is a triangle. The sorting mechanism I use is somewhat different. During a break in the meeting, the facilitator copies all of the sentences and ideas listed on the board to a piece of paper and makes photocopies for each participant. When the meeting resumes, each person is asked to label each statement as either a Focus item or a Not Considered Now (NCN) item.

Statements labeled as NCNs can include items that we will accept as fact. For example, someone may list "We must realize returns in excess of our costs." Although this is a good guiding principle, it likely will not be part of the expression of the final question. NCN items can also include details. These are items that may eventually be addressed, but not as part of the primary problem. For example, someone may state "We must train our people in food-processing regulations." This may be true if we decide to implement the food-processing strategy; however, if we do not implement that strategy, training is not necessary. Statements labeled as "focus" items are those that we think are the important problems upon which we should focus our attention. These are the statements that will eventually form the essence of the problem to be solved.

After each person has labeled all statements with one of the two labels, the papers are gathered and those statements that were identified as "focus" statements by more than about half the participants are printed on a new paper and again distributed to the group. If there are relatively few statements that were identified multiple times as focus items, then it is expedient to resort to the common-element process described previously. That is, can we express the elements of the few focus statements as a single problem to which (nearly) everyone can ascribe? If so, then the statement of the problem should be generated from the few focus items. If, however, the number of focus items is great or diverse, then a second round of labeling only the statements previously identified as focus items is undertaken. This time, the people in the group are asked to sharpen their focus and again label each item as either an NCN or a focus item.

In particularly difficult situations, several iterations of this process may be required to arrive at a small set of statements. The small set of statements should either be reduced to a single expression of the problem by identifying common elements in the statements or be considered to be separate problems if no common ground can be found. In the latter case, the facilitator should make the group aware that separate risk assessment models may be required for each statement and that the group will have to decide in which order the individual problems will be addressed.

GROUP DYNAMICS AGAIN

Sorting schemes such as that described here not only are valuable tools for arriving at consensus with regard to just what the problem is, but are also great tools for group dynamics. When an individual in a group contributes an idea or statement, that person would like to feel that his/her contribution has been given a fair hearing and consideration. Using these sorting techniques, the contributed ideas that do not seem to apply need not be singled out as inappropriate.

The sorting process described will discard less useful contributions along with many others of greater merit. In this way, the feelings of people who have contributed tangentially useful ideas are not singled out, nor do they feel criticized for their contributions. It is an important part of the facilitator's job to keep as many people contributing to the process as long as possible. Tools such as this sorting mechanism help very much in this regard.

UNITS

Following succinct definition of the problem to be solved, the next item to be decided is just how, in the end, the answer will be expressed. This may seem like a technical detail, but it is not. It will guide the rest of the entire risk assessment exercise.

Typically, if the question is well defined, a discussion of units will follow. For example, if we are considering the problem of adding production capacity to an existing plant, we may wish to express the final answer in tons of product, or throughput of material, or a financial measure such as cash flow, or after-tax income, or net present value, or any of a host of other financial measures. In any event, the units of the answer must be decided upon at the beginning of any risk assessment.

In the next section, I discuss in detail what I call "overarching categories." Examples of such categories may be Economic, Environmental, or Technical. Each of these categories typically is broken down into numerous variables. For example, categories and variables might be

- Economic
  — Price (dollars-per-gallon) of product
  — Advertising costs
  — Litigation costs
  — Construction costs
  — Tax rate
  — Operating expenses
- Environmental
  — Probability of a spill
  — Attitude of local government toward pipeline construction
  — Previously existing environmental damage
  — Availability of emergency cleanup equipment
- Technical
  — Capacity of new pipeline
  — Capacity of existing facilities

It is important to establish the units of the answer, because all variables that contribute to the solution of that answer must, in the end, be able to be combined so that the result is in the desired units. For example, the answer variable may be "after-tax income in dollars." In this case, we must be able to combine in our risk model variables such as Probability of a spill, Attitude of local government toward pipeline construction, and all other variables in such a way as to result in cash flow in dollars.

So, knowing that the end result of the analysis must be expressed in dollars, we must keep in mind when discussing the Probability of a spill variable that the probability has to be multiplied by some dollar amount. We may decide to integrate the variable Attitude of local government toward pipeline construction into the risk model by asking questions such as: "What is the probability that the local government will stop the project altogether?" and "What is the per-week cost of delay if the local government holds up construction?" This variable might also be handled by implementing a unitless scale that can be translated into dollars. For example, we may wish to express the attitude of local government toward such projects on a scale of 1 to 10 with 1 indicating a "good" attitude toward such projects and 10 indicating a "bad" attitude. You may then in your risk model use this 1-to-10 value as a multiplier (or as a pointer to a multiplier) for a value or distribution of values which represent, for example, the weekly cost of local government delays.

It should be clear from the examples listed here that the definition of the output or answer variable and the units in which it will be described are important. This must be known prior to defining categories and variables for the risk model. Certainly it must be taken into consideration when phrasing questions that will be used to prompt information providers.

OVERARCHING CATEGORIES

Every problem is composed of component parts. In the risk business we generally think of these component parts as variables. We do this even though some of the components, in the end, will not vary (constant values or chances of failure, etc.). Thus, I use the term *variable* here in the loosest sense.

The variables that comprise a risk model generally can be categorized. In constructing a risk model, I generally approach the situation by first defining what I call "overarching categories" (OCs). Later I help the group generate the variables that collectively define each category.

For example, in the case of a plant-construction risk model, I would next challenge the group to define categories for consideration in the risk model. When building a plant, certainly you would have to consider aspects in at least the following categories

- Political
- Environmental
- Labor (union)
- Technical
- Commercial
- Financial

When making the list of categories, be sure to generate a definition for each category. For example, when we list Labor as a category, just what do we mean? If we are going to consider only the cost of labor, then why would that not be considered under the Financial category? Considerations such as the availability of qualified labor, negotiations with local unions, and other labor-related considerations may justify Labor as a category separate from the Financial one.

Categories are useful for organizing variables and for orchestrating discussion of how the risk model will be constructed. In addition, categories help to organize presentation of the risk model to the end user. When risk models are constructed, variables typically are added to the user interface in an order that generally is dictated by their appearance in equations or by the order in which they arose in discussion of the model. Categorizing the variables allows for a logical grouping and flow of information in the model user interface.

13 Build a Contributing-Factor Diagram

CONTENTS

THE CONTRIBUTING-FACTOR DIAGRAM — GETTING STARTED

As the risk model takes shape, variables that contribute to the solution of the problem will be added. Names for these variables should be agreed upon and the names recorded on some type of movable object such as a large Post-it® note or a magnetic "octagon" (the type used in systems-thinking exercises). These name-bearing objects should have on them sufficient space for recording of the name and some additional information. Post-it® notes are relatively inexpensive and can be "stuck" on a vertical board for viewing and arranging. These arranged items can comprise a contributing-factor diagram (CFD). An example of a CFD is shown in Figure 13.1.

I am an advocate of the CFD. It is an outgrowth of the more formal and rigorously constructed "influence diagram." In the classical and academically adherent influence diagram, a mix of symbols is used to represent nodes of various types. Years ago I used to try to construct influence diagrams to outline the system of equations for a risk model. However, during the initial stages of a project I knew so little about how the various components would, in the end, interact that it was foolish to try. Even now, after many years of experience in building such diagrams, I find I am doing really well if I can construct a rudimentary arrangement of the fundamental project components.

Before influence-diagram huggers renounce me as a risk-heretic, I would here like to adduce that influence diagrams certainly do have a prominent role in modeling. Such diagrams can be used for at least two purposes. One common application is to use the plots to outline a work process. Another application is to utilize the influence diagram to delineate the logical and mathematical relationships between

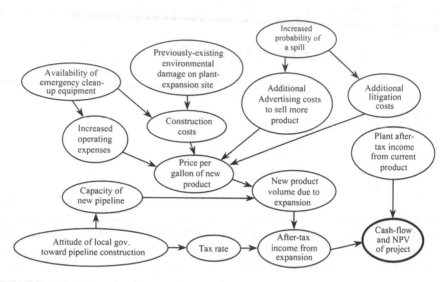

FIGURE 13.1 Example of a contributing-factor diagram.

risk-model variables. It is for this latter application that I recommend the simpler CFD. A detailed plant-construction example will be given later in this chapter.

Something to realize about a CFD is that it is not a flow chart. In a flow chart, entities are arranged in sequence. Things that happen first generally are arranged at the top of the flow chart, and items that are subsequent are listed below. In contrast, the items arranged on a CFD are not time- or sequence-dependent. Items are connected with arrows that indicate that one item is used in the calculation or determination of another. For example, in Figure 13.1, the variable Probability of spill is used in the calculation of Advertising costs and Litigation costs. The fact that these variables are posted near the top of the diagram has nothing to do with whether, in the risk model, they are used prior to or after any of the other variables shown.

In a CFD we generally work backward from the answer. That is, having previously succinctly defined the question and the answer variable and its units, we post the answer variable first on the CFD. Before proceeding with the posting of additional variables, however, there is one more answer-variable attribute to be discussed. The next step in CFD construction is to decide just how the answer to that question will be presented. In Figure 13.1, the answer variable is After-tax income. We will express this in dollars. At the time of CFD construction, the group must come to consensus concerning just how the output variable will be displayed and presented to management. Later in this book, it will be disclosed just what a distribution is; however, it is important at this juncture to consider some practical aspects of distributions.

In Figure 13.2, the variable After-tax income is displayed by three different bar charts. The data making up each bar chart are identical; the only difference is the number of bars we selected into which the data would be subdivided for display. It can be seen that when only 5 bars are selected, the bar chart has a roughly lognormal

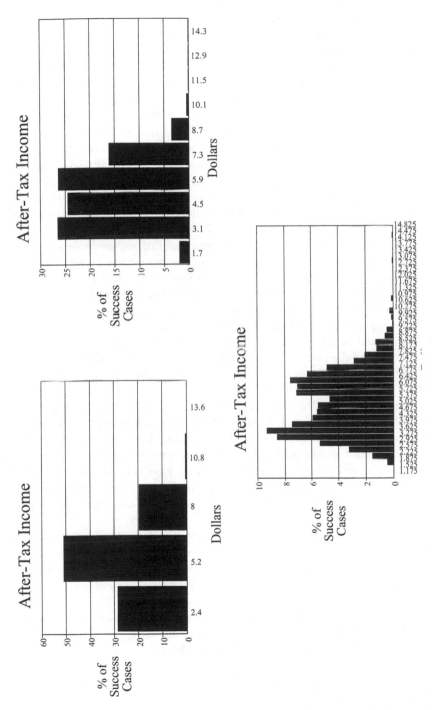

FIGURE 13.2 Frequency plot of common data using 5, 10, and 40 bins.

appearance (skewed to the right). When 10 bars are selected, the lognormal character is lost; it appears that there is a part of the distribution where the frequencies are all similar and that there are outlying values on either side of this range. When 40 bars are selected, the bar chart takes on a bimodal appearance. These three bar charts of the same data distribution certainly would convey different messages to those not well grounded in statistics.

The three different "pictures," however, are not the worst of it. Consider the situation in which project economics dictate that if the project cannot generate about $5 million, then it is likely it will not be pursued. In the 5-bar chart, the probability of generating "around" $5 million is about 50% contrasted with the 10-bar probability of only around 25%. The 40-bar chart indicates the probability is less than 9%. It should be obvious to those versed in statistics that the differences in probabilities are related directly to "bin" sizes. Managers to whom the bar chart of After-tax income is presented, however, may not be so savvy in statistical inference and certainly will not be presented with multiple plots of the same data for comparison. This topic is addressed in greater detail in Chapter 15.

At the time of CFD construction, it is essential that the group decide just how much detail (how large or small the "bins" need to be) is to be conveyed in the bar chart. Group members should come to consensus on this issue and record exactly why they decided on a particular level of resolution. All members of the group should agree to present the After-tax income data consistently.

Following resolution of the problem of how answers will be presented, additional variables should be added to the CFD. These additional variables should be those that contribute to the calculation of the answer variable and which have passed the tests outlined in the Identify and Define Variables, Ask the Right Question, and Double Dipping sections of this chapter.

For each variable added, the group should agree on a written definition of the variable and exactly what will be asked of the user. For example, in Figure 13.1, one of the variables listed on the initial contributing-factor diagram is Attitude of local government toward pipeline construction. At first blush this may seem like the proper question to ask. Consider, though, that we have defined our output variable to be After-tax income. Therefore, we must gather from the user information relating to local government attitude that will be included in a calculation that results in After-tax income expressed in dollars.

Upon further consideration, we may decide that Attitude of local government toward pipeline construction is not exactly the question we need to pose to supply data to the risk assessment model. In this case we may want to decompose the original question into the following set of questions

- What is the probability that local government will delay pipeline construction?
- What is likely to be the length of time of any delay?
- What is the cost per unit of time of the delay?

Users would be prompted to supply the answer to the first question as a distribution of probabilities expressed in percent. Units for the second question related to time may be a distribution of months. A distribution of dollars-per-month would suffice as input for the third question. The user-supplied data then could be combined as

(probability of delay) × (length of delay in months) × (cost per month)   (13.1)

In this way, political considerations can be used to contribute to the calculation of After-tax income.

In addition to generally posing each consideration so that it can be used in the calculation of the identified output variable, it is essential that the group agree upon the exact wording of the question/statement that will prompt the user to enter the desired data. For example, for the first question listed above, the group might settle on the phrasing:

> Enter below a distribution of decimal percentages that represents the probability that local government will delay the project due to its attitude toward pipeline construction in the project area. This probability should not include consideration that the government will cancel the project altogether, only that the local authorities will delay the project.

Wording of the question/statement is critical. In this instruction, it is clear that probabilities-of-delay related only to the government's attitude toward pipeline construction in the project area should be considered. This should not include, therefore, consideration of other reasons the company might have for delaying the project (lack of funds, availability of materials, etc.). In addition, the probability that the government will cancel the project altogether may need to be acquired separately. This is a different question relative to calculating the cost of a government-imposed delay. From this simple example it should be clear that the wording and phrasing of questions are critical to gathering the desired information from users.

Pertinent variables are added one at a time to an increasingly detailed and evolving CFD. As each is added, it is placed on the diagram in a position that indicates its relationship to variables already existing on the diagram. Arrows are drawn from the newly added variable to variables that use the newly added variable in their calculation or determination. A poorly designed CFD can appear like a plate of spaghetti with intersecting and crossing arrows. In a well-designed diagram it is easy to trace factors backward from the answer variable to the most important input variables. A well-designed CFD is a blueprint for generating the equations that will form the basis for the risk assessment model.

IDENTIFY AND DEFINE VARIABLES

Following the establishment of categories and the posting of the answer variable on the CFD, the next step is the identification of variables that make up the categories and the establishment of the relationships between the variables. The risk-process facilitator should guide the group through a discussion of each of the categories to establish just what variables might be appropriate. As outlined in Chapter 12 at least the following categories and variables might be established for the plant-construction project.

- Economic
  - Price (dollars-per-gallon) of product
  - Advertising costs
  - Litigation costs
  - Construction costs
  - Tax rate
  - Operating expenses.
- Environmental
  - Probability of a spill
  - Attitude of local government toward pipeline construction
  - Previously existing environmental damage
  - Availability of emergency cleanup equipment
- Technical
  - Capacity of new pipeline
  - Capacity of existing facilities

Just as categories can be broken down into sets of variables, the variables themselves can be subdivided. For example, it may be necessary to subdivide the construction cost variable into component variables such as heavy equipment rental costs and compressor (purchase) costs.

It is the job of the facilitator to ensure that any given subject or variable is not unnecessarily subdivided. Too much detail is detrimental, so the group should strive to keep variables as general and "high level" as possible. It also is essential that any question devised to query a user for variable-related information be posed so as to prompt the user to provide information that will best serve the purpose of the risk assessment.

ASK THE RIGHT QUESTION

Asking the right questions (in addition to correctly phrasing them) can be critical to success. For example, one consulting assignment with a group charged with supplying a client with risk assessments for construction costs encountered a fundamental problem. The actual project costs deviated significantly from those predicted by the risk assessment model run prior to project inception. Analysis of the risk assessment process employed by this group revealed several problems, not the least of which was the phrasing of questions that prompted users to supply information. Some of the questions posed did not contribute to the solution of the actual problem.

The ultimate goal of the group's risk model was to predict, prior to beginning the project, the total project costs. One cost was compressor purchases. Because the risk model was intended to be run prior to the project's actual beginning, the risk model asked, among other questions

- How many compressors will be needed?
- How much does a compressor cost?

The user generally entered a single value (say, 1) for the answer to the first question and entered a distribution of costs (say, $100,000 plus or minus 10%) for the answer to the second question. Because these costs and the calculated total costs were being generated prior to the project start, the project could still have been rejected because it was absolutely too costly or because it cost more than a competing project. For these reasons, there was considerable pressure by project proponents to minimize costs. The costs for compressors are fixed by manufacturers, so there was little room for cost savings there. However, it could be proposed that a small number of compressors would be needed, thus keeping costs down.

The actual number of compressors needed was generally more than the number entered in the preproject risk assessment model. Therefore, it should have come as no surprise that the actual total cost of the project significantly exceeded initial preproject estimates. To remedy this problem (in part), I suggested that they ask different questions with respect to compressor costs. The questions I proposed they ask are

- When projects of this type are finished, how many compressors did we typically use?
- How much does a compressor cost?

Asking these questions will yield much more accurate total-project costs, with respect to compressors, than would the original set of questions. Much of a facilitator's job centers around limiting the amount of detail (yet capturing sufficient detail) demanded by the model. In addition, the facilitator should be concerned with asking the right questions. Pertinent questions should be phrased so that answers to questions result in user-supplied information that actually contributes to answering the question that is the aim of the risk assessment model.

DOUBLE DIPPING

Generally, the facilitator is much less familiar with the problem than are the members of the group. However, a facilitator should use common sense and his or her relative ignorance to advantage. Realizing that no one else in the room can ask "fundamental" questions with impunity, the facilitator should question the addition of each variable and consider whether the information to be captured by the new variable is not, in fact, already captured by some other variable or set of variables. Avoidance of "double dipping" (capturing essentially the same information more than once, but perhaps in different ways) always is a major concern in model building. Often, group participants are "too close to the problem" to realize some of the overlaps in logic. This is especially true when representatives from different disciplines approach a common problem.

For example, it may be deemed that operating expenses are too high. An engineer may attempt to reduce operating expenses by suggesting that certain types of more expensive but more durable materials be used in the operation, thus lowering operating expenses in the long term. The economist, however, has already mentally discounted the use of those materials because they are too costly and proposed adding a third shift to reduce overtime expenses. Both approaches solve a common problem (lowering operating expenses). However, we would like to keep construction costs down (use cheaper materials if possible), and we would like to keep the number of employees to a minimum (two shifts better than three). Thus, we would likely invoke one, but not both, of these operating cash-reduction methods. It generally falls to the risk-process facilitator to uncover the duplications. Getting each participant to express *why* he or she is suggesting a particular approach for solution to a problem and employing experience and common sense are typical ways to guard against double dipping.

DOUBLE DIPPING AND COUNTING THE CHICKENS

Double dipping appears in many insidious forms and is a difficult hazard to avoid when building a model. A good facilitator attempts to ensure that no piece of information is collected more than once in a risk model. Avoidance of this problem is, more or less, controlled by the facilitator and the model builders. However, once the model is built and "shoved out the door" to the user community, double dipping of a different type can take place. A particularly menacing one I call the "counting-the-chickens" problem.

Only organizational and process-related safeguards can even begin to avoid the counting-the-chickens problem. Consider that we have five houses that form a circle and, therefore, share a circular back yard. In the yard there are a dozen chickens. We knock on the front door of each house and ask the occupants how many chickens they have. They look out the back window and report that they have 12. This same answer is given at every house, leading us to believe that 60 chickens exist. Such counting of things more than once is a common failing in the organizational implementation of a risk assessment process.

FIXING THE DOUBLE DIPPING AND COUNTING-THE-CHICKENS PROBLEM

The counting-the-chickens problem is an especially rancorous dilemma to solve. There exists, to the best of my knowledge, no technical basis for its solution. Organizational awareness and aplomb tend to be the only remedies. Organizational cures, although potential solutions, are not easily implemented.

Consider a company that produces a product in several locations around the world. At each location the company expects that the lion's share of the product will be consumed by the local indigenous population. The volume of product produced in excess of the amount consumed locally is expected to be exported to a targeted population of consumers. It is not uncommon to find that multiple production facilities have based their business plans on a common population of consumers. Those consumers might be able to absorb the excess production from one facility, but certainly not from all plants. In this situation, how do we prevent the counting-the-chickens problem (i.e., counting the same consumers more than once)?

Organizational processes are probably the only avenues to a possible remedy, but organizational solutions can be notoriously untenable. As an example, the aforementioned corporation might decide that it will now require that all such proposals be funneled through corporate headquarters prior to actually being implemented. At corporate headquarters, then, it must be somebody's job to investigate each proposed production plan and to uncover any overlaps in the plans.

This solution works well when the data from the producing locations are honest and accurate. Exporting is more expensive than local consumption, so the plans may "play down" the amount of exports. This can lead the corporate overseer to believe that the consumers targeted to absorb exports are capable of doing so for more production facilities than they actually can. In addition, the corporate overseer position often is not a popular one. It does not tend to be a career-path position and is viewed with some degree of resentment by the operating facilities that are being "governed." Therefore, this position typically experiences high turnover; that is, those assigned to the job generally take the first opportunity to vacate the position. Without consistency of staffing in this critical assignment, much of the effectiveness of sleuthing out problem areas is lost.

Another popular remedy is to assign a person or group to temporarily become part of the production-site team that is going to put forward a proposal. The assigned person generally "moves" to the production site and becomes part of the process and organization that will generate the proposal. The idea here is that such a person who joins multiple teams in different locations will be able to spot the problems before they become part of the proposal. This approach, however, has some obvious drawbacks.

First, the employees whose job it is to join the location-specific team must be willing to move to the remote site for an extended period of time (and to repeat the process when the current assignment has ended). This can definitely be done, but it cannot be done indefinitely. In addition, more than one proposal might be fabricated at the same or different facilities. This requires that there be more than one journeyman. If more than one person is involved, then avoiding the counting-the-chickens problem depends upon close coordination and good communication between those individuals. This, in itself, can be difficult to implement.

Although both solutions outlined here are only partial cures and are far from perfect, implementing some manner of organizational control to avoid double dipping is better than instituting no plan at all. The counting-the-chickens problem is a particularly insidious risk-modeling problem and can wreak havoc on a company if not suitably addressed.

CFD-BUILDING EXAMPLE

Let us consider a situation in which Joe has to build a new chemical plant. In the initial planning stages, Joe might wish to construct an influence diagram that will depict the stages and steps that need to be taken between the present time and the opening of the plant. In such an influence diagram, Joe would begin by posting the "answer" variable — the opening of the plant. Joe knows that at least the following variables will influence the plant opening, and, therefore, must be included in his diagram.

- Negotiations and contract signing with the labor union (negotiations may not be successful)
- Identifying local suppliers for building materials (there is no guarantee that critical supplies can be had locally)
- Governmental approvals and permits must be obtained, but permits have, in the past, been denied for similar projects
- Engineers must submit a construction plan that not only is economical, but also will comply with environmental regulations and will result in a plant that will fit on the site and provide sufficient production capacity (this is not a given)
- Local environmental groups must be placated

Joe will commence his project by organizing the aforementioned items and others as an influence diagram (not shown here). Rectangles (decision nodes), ovals (uncertainty nodes), and octagons (value nodes) will be used to represent the individual items. For this type of planning, influence diagrams are invaluable.

Ultimately though, Joe will be faced with the arduous task of calculating, for example, the net present value (NPV) and perhaps the internal rate of return (IRR) of the plant. In the initial stages of the project, Joe does not have a firm grip on exactly what variables will contribute precisely to a set of equations. He does have some inkling as to the general relationships between the variables that will be used to construct the strict set of equations of the risk model. For this purpose, at this stage, a CFD is more suitable.

In a typical CFD, all variables are represented by a single common symbol — an oval, a rectangle, or the like. No intrinsic meaning is attached to the symbol. Symbols are, like those in an influence diagram, connected by arcs and arrows to indicate which variables contribute to the calculation of other components (even though, at this stage, the precise equations that might be used to calculate a given variable might not yet have been decided). An initial stage of a CFD is shown in Figure 13.3.

In this figure, Joe has posted the "answer" variable (NPV and IRR of the new plant) and several major factors that will contribute to the calculation of NPV and IRR. Joe knows that he will have to calculate Total costs and that, at least in the initial stages of the project, Construction costs will be a major part of the Total costs. He knows also that to have healthy NPV and IRR values, he must more than offset the costs by generating sufficient income from the plant.

In a meeting with a facilitator and all other pertinent players in the plant-construction project, Joe presents his initial CFD and asks for comments, additions, etc. One of the first things to be pointed out is that the Construction costs variable needs to be subdivided into its major component parts. The facilitator, in an attempt to keep things simple, challenges the breakdown of the Construction Costs parameter but is quickly convinced that at least some subdivision is pertinent.

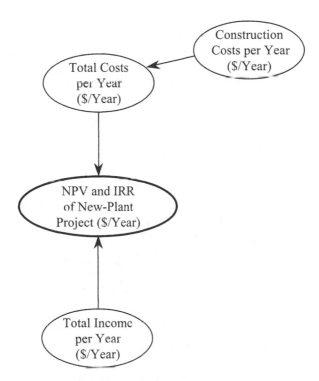

FIGURE 13.3 Initial contributing-factor diagram.

Subdivision proponents argue that if the plant project is killed in the initial stages by abnormally high construction costs, it will be necessary to know what facet of those costs — labor or steel or concrete or whatever parameter (or combination of parameters) — caused the Construction costs variable to contain coefficients of unreasonable magnitude. As shown in Figure 13.4, the original Construction costs variable will comprise the combination of costs related to labor, pipe, steel, compressors, concrete, and exchangers.

Labor, it is decided, will be considered as two types. Construction labor is considered a capital expense (CAPEX) while "operating" labor (plant workers, office staff, etc.) is considered an operating expense (OPEX). Labor will be calculated from the factors Days per year, Hours per day, Hourly wage, and Number of workers. This will be done regardless of whether Joe is calculating CAPEX labor costs (represented by the arrow between Labor costs and Construction costs — Yearly in Figure 13.4) or calculating OPEX labor costs (represented by the arrow between Labor costs and Total costs per year in the same figure).

Joe's group decides that Days per year not only contributes to the estimation of Labor Costs, but also will serve as a term in the equation that calculates Daily plant income. The Dollars per ton and Tons per day variables also will serve as terms in the equation that calculates Daily plant income.

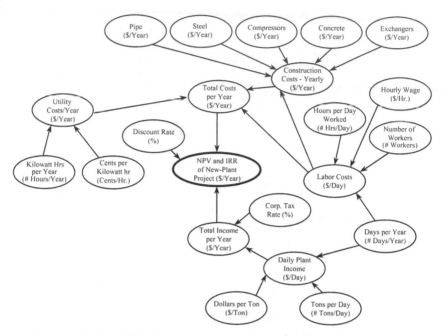

FIGURE 13.4 Completed contributing-factor diagram.

Variables Daily plant income and Corporate tax rate will be the major contributing factors in the calculation of Total income per year. Along with Discount rate, this answer parameter, represented as a time-series variable, will, in turn, be the major contributing factor in calculating yearly cash flow for the NPV and IRR estimates.

In addition to Labor cost OPEX and Construction costs, Joe's team decides that the other major contributing factor in the calculation of Total costs is Utility costs per year. At the time of the initial CFD-building meeting, Joe's team decides that two factors should contribute to the calculation of Utility costs per year. Kilowatt hours per year and Cents per kilowatt hour are added for this purpose. However, it will be discovered later that the team decides that if utility costs are exorbitant, there is no real information, all things considered, in knowing whether the number of hours or the cents per hour is the culprit. In addition, the team, through use of the model, will discover that users are more adept at supplying total energy costs per year and less adept at providing accurate and precise information about hours and cents.

The team decides that the Construction costs should be treated in a separate model. The output distribution from the Construction costs model will be used as an input distribution for the main NPV/IRR-calculating model. The team decides this because the construction phase of the plant is imminent and the construction teams, at this stage, have to rerun the model multiple times in an attempt to emulate several construction scenarios. In addition, this cadre of construction consultants has a pretty good grip on the associated costs, but they do not have a clue concerning items such as Discount rate, OPEX costs for labor, Dollars per ton, and some of the other variables. They would be obligated to supply values for such variables if the model is not divided.

One more point to note about the CFD in Figure 13.4 is that each variable, starting with the answer variable (NPV/IRR), is labeled with units. A good facilitator will preclude adding to the CFD (not an influence diagram) any variable that cannot be put into a set of equations and somehow used to calculate the answer variable. This is an important point because the risk-model builder likely will not have all team members available at the time of building the model and will have to rely on the CFD as a blueprint for model assemblage.

SHORT LIST OF HINTS FOR BUILDING A CFD

The numbered steps below are a synopsis of how to approach the building of a CFD.

1. Decide what the "answer variable" is to be (this can be difficult). There does not have to be just one "answer variable", but decide on just one to start with.
2. Decide in what units you want the answer variable to be expressed. For the sake of this example, let us consider that you want to calculate NPV and that you want it expressed in dollars.
3. Begin to add large variables to the diagram that would contribute to NPV, such as Costs and Incomes. Then, for each large variable, show the things that go into making up the large variable. For example, Costs might consist of two variables called OPEX and CAPEX. OPEX might consist of Labor costs, Rental fees, etc. CAPEX might be composed of the Cost of steel, the Cost of land, etc. Likewise, Incomes might include Number of barrels sold, Price per barrel, etc.
4. For each variable that you add, be *sure* to decide in what units it would be expressed. For example, if you add a variable to the diagram called Politics because you think there might be political problems with the project, consider that your output variable (NPV) is expressed in dollars. How would you include Politics in an equation that would result in dollars? It might be better to change the Politics variable to three new variables.

   - The probability (percent chance) that we will have a political problem (expressed as a distribution of values between 0 and 1)
   - The length of time we might be delayed if we have a political problem (expressed as a distribution of days)
   - The dollars per day it will cost us if delayed (expressed as a distribution of dollars)

Now the percent probability of delay, the length of delay, and the dollars per day can be multiplied to obtain a dollar value that can be used in an equation to calculate NPV.

Likewise, do not put a variable on the contributing-factor diagram that says, for example, Electricity. What you really want is Kilowatt hours expressed as a distribution of Hours and Cents per kilowatt hour expressed as a distribution of cents. These can be combined to influence NPV (in dollars).

If you use the logic outlined above, you should end up with a contributing-factor diagram on which each variable can contribute to an equation or set of equations that will, eventually, calculate cash flows and then NPV.

14 Monte Carlo Analysis

CONTENTS

A BIT OF HISTORY

Monte Carlo analysis uses the process of simulation to achieve a range of solutions to a problem. This technique generally is used to solve problems for which the definition of specific solution equations to calculate a specific answer is either too complex or too cumbersome to be practical. In Monte Carlo analysis, input variables are typically (but not necessarily) represented as distributions of values (see Chapter 15). Values making up these distributions represent the range and frequency of occurrence of the possible values for the variables. Also needed are equations that relate the input variables to the output (result or answer) variables. Monte Carlo analysis is a statistical procedure based upon random selection or chance. The name, of course, is taken from the city of Monte Carlo, Monaco, made famous by games of chance.

In the early 1900s, Monte Carlo analysis, or distribution sampling, arguably first was presented formally under the guise of "Student" (W. S. Gosset, inventor of the Student's t-test). At first, Monte Carlo analysis was treated as being distinct from "experimental sampling," or "model sampling," both of which involve random selection from populations. The term *Monte Carlo* was restricted to processes in which weighted or importance sampling was used to increase the efficiency of the sampling process. In other words, the term *Monte Carlo* was reserved for sampling processes that embodied some "smarts" to increase sampling efficiency. This restricted use of the term eventually faded because high-speed computers lessened the importance of an efficient sampling process.

Today, the term *Monte Carlo analysis* can be applied to any procedure that uses distribution-based random sampling to approximate solutions to probabilistic or deterministic problems. These types of problems were exemplified by the first practical use of the Monte Carlo method — addressing the probabilistic aspects of particle diffusion associated with the design and construction of the first atomic weapons at Los Alamos during World War II.

FOR WHAT IS IT GOOD?

Most processes — physical, chemical, or otherwise — terminate in a result. Many processes are so complex that the generation of an equation or system of equations to calculate their results is arduous, if not impossible. Monte Carlo analysis often is used to simulate these processes and to generate probable results. This avoids having to resort to the construction of specific equations which can predict the outcome of the process. In addition, Monte Carlo analysis can yield likelihoods for a range of possible results.

Although Monte Carlo analysis has many uses in business and science, the most common application involves determining the probability that a certain event (or result) will occur and predicting the magnitude of the event. The fundamental technique used to accomplish these is the random sampling of values from frequency distributions which represent the variables in the analysis (see Figures 14.1 and 14.2). But, what does all that mean?

Every risk assessment study involves the integration of at least several, and sometimes many, variables. Variables are the basic elements of the problem (porosity of the soil, concentration of an element in groundwater, distance from a potable water source, etc.). If Monte Carlo analysis is to be used, each of these variables is usually represented by a distribution of values (coefficients of the variable). The values that constitute the distribution should span the range of possible values for the variable in question.

Monte Carlo analysis, in its native form, also requires that the variables be independent. That is, they must not have a relatively high correlation. For example, if two variables in a Monte Carlo analysis are depth and temperature, then it is not valid to randomly select a depth and temperature value from their respective distributions and expect that this combination will emulate a real situation. This is because, in general, as depth increases, temperature increases. A randomly selected depth of 5 ft and a temperature of 500°F do not, in most cases, simulate real-world conditions. To account for the correlation between variables, the technique of defining dependence between the two variables should be employed. See Chapter 17 for an explanation of this technology.

SIMPLE MONTE CARLO EXAMPLE

For the sake of an example of Monte Carlo analysis, let us consider the oversimplified example of calculating profit from income and payments (payments = salaries, taxes, and other costs). Figure 14.1 shows a distribution of possible incomes for the year. Figure 14.2 shows the range of payments we make in the same year. On each iteration of Monte Carlo analysis, a randomly selected income will be combined with a randomly selected payment to yield a profit value using, for example, the equation

$$\text{profit} = \text{income} - \text{payment} \tag{14.1}$$

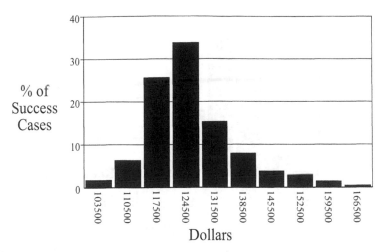

FIGURE 14.1   Distribution of income.

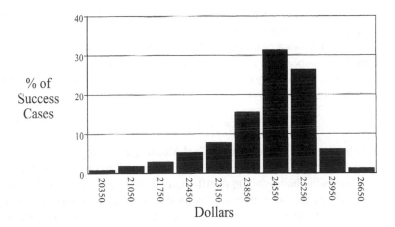

FIGURE 14.2   Distribution of payments.

The plot of profits shown in Figure 14.3 is the result of 1500 random selections and profit calculations. This model, of course, assumes that there is no relationship between income and payments. This is not likely. See Chapter 17 for a discussion of dependent relationships between model input variables.

HOW MANY RANDOM COMPARISONS ARE ENOUGH?

The object in Monte Carlo analysis is to perform a sufficient number of comparisons from the input variable distributions so that the resulting variable distribution is valid. Some risk assessment packages allow the user to specify a convergence value which will automatically stop iterations when comparisons of mean values (or other selected values) from multiple "simulations" is less than or equal to the assigned convergence value. However, the number of distributions, the complexity of the

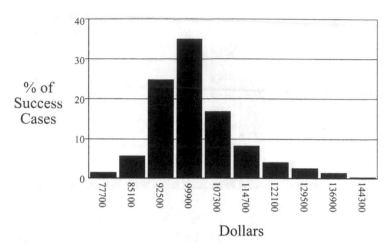

FIGURE 14.3 Distribution of profit resulting from Monte Carlo combination of income and payments.

equations, and the familiarity of the user with the situation may preclude knowing such a convergence value prior to the analysis.

A simple method for determining whether a sufficient number of comparisons has been made is to inspect the mean value of the "answer variable" distribution from multiple Monte Carlo analyses. For example, if we have two input variables called X and Y and our equation is

$$Z = X + Y \qquad (14.2)$$

then the Z distributions from multiple individual Monte Carlo analyses (an individual Monte Carlo analysis may be comprised of, say, 5000 iterations) would be inspected. If the means of the individual resulting distributions do not vary significantly (significance is determined by the user), then it is likely that a sufficient number of comparisons has been made. Statistical tests for the significance of the difference of population means (t-tests) or of the entire distribution using the Kolmogorov–Smirnov test are available but can be complicated. Most of the time, simple inspection of the change in the mean of the resultant distributions and the shape of the distributions will suffice.

OUTPUT FROM MONTE CARLO ANALYSIS — THE FREQUENCY AND CUMULATIVE FREQUENCY PLOTS

As depicted in Figure 14.3, the primary output vehicle for Monte Carlo analysis is the frequency plot. The number of bars in a frequency plot can be set by the person generating the model. Bar heights simply indicate the number (or percent) of X-axis values that fall within the X-axis range dictated by the width of the bars.

Although useful, frequency plots are limited in the amount and type of information conveyed. In risk assessment and portfolio management, it is more common to view Monte Carlo output in cumulative frequency space. The same data shown

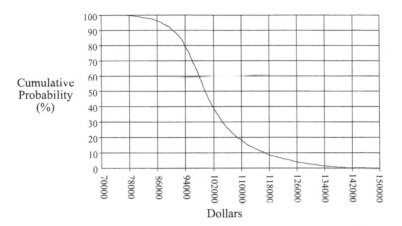

FIGURE 14.4 Cumulative frequency plot equivalent to frequency plot in Figure 14.3.

in frequency space in Figure 14.3 is depicted in cumulative frequency space in Figure 14.4.

The conversion of frequency space to cumulative frequency space can be conceptually viewed as "stacking" the bars of a frequency plot. Figure 14.5 shows a four-bar frequency plot. In Figure 14.6, the same four bars are "stacked" from right to left. In Figure 14.7, a dot has been inserted at the top of each bar and the dots connected with a line. In Figure 14.8, the bars have been removed, leaving only the cumulative frequency curve.

In the computer, of course, there actually are no bars. Software systems that build cumulative frequency curves work directly with the individual X-axis values generated by Monte Carlo analysis. Figure 14.9 shows a plot of eight values resulting from an eight-iteration Monte Carlo risk assessment model. The eight values have been sorted from "smallest" (values of least magnitude) on the left to "largest" (values of greatest magnitude) on the right.

Because eight values were generated, the cumulative frequency algorithm will divide the Y axis into eight equal probability segments. The "largest" value is moved up one probability segment. The second largest value is moved up two probability segments and so on until the algorithm reaches the last data point (leftmost or "smallest" X-axis value) which is moved up, in this case, eight probability segments (Figure 14.10). The points are connected with a smooth line which is the cumulative frequency curve (Figure 14.11).

INTERPRETING CUMULATIVE FREQUENCY PLOTS

Now that we have constructed a cumulative frequency curve, what good is it and how do we read one? Frequency can be accumulated to the left or to the right, and the resulting Z or S curves, respectively, impart similar information. However, there is a slight difference in the interpretation of the two curves.

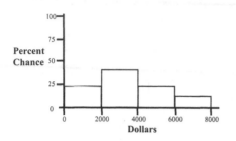

FIGURE 14.5 Four-bar frequency plot

FIGURE 14.6 Bars from Figure 14.5 stacked.

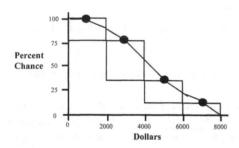

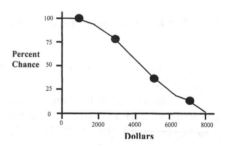

FIGURE 14.7 Curve connecting tops of bars from Figure 14.6.

FIGURE 14.8 Cumulative frequency plot equivalent to frequency plot in Figure 14.5.

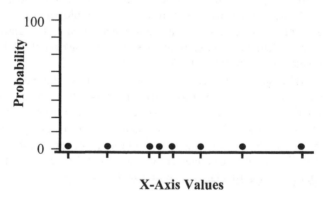

FIGURE 14.9 Results from an eight-iteration Monte Carlo analysis.

Both curve types are shown in Figures 14.12 and 14.13. The X axis for both curves is profit in dollars; the Y axis is probability. The dashed lines on each plot connect a dollar amount with a probability. From Figure 14.12 it can be interpreted that there is a 42% chance of attaining a profit of about $500 *or greater*. Or, stated

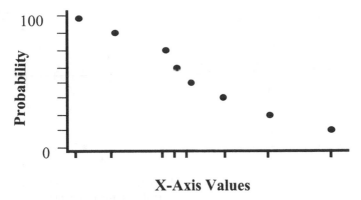

FIGURE 14.10 Individual *X*-axis values from Figure 14.9 "raised parallel to the *Y* axis according to their cumulative probability.

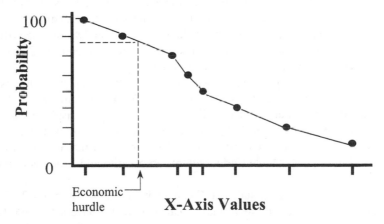

FIGURE 14.11 A cumulative frequency curve constructed by connecting the points in Figure 14.10.

differently, there is a 42% chance of getting *at least* $500 of profit. Figure 14.13 can be interpreted to indicate that there is a 40% chance of getting *up to* a profit of about $400. Stated another way, there is a 40% chance of obtaining a profit of $400 *or less*.

People who deal with situations where bigger is better prefer the "accumulating-to-the-left" (Z-shaped) plot. For example, a petroleum engineer charged with esti-mating oil production from an oil well would want to know that the well will produce X amount of oil or more (or, at least X amount of oil). A cost accountant, on the other hand, might prefer the "accumulating-to-the-right" curve (S-shaped) because in her job it is important to know that a project has a Y-percent chance of costing "up to" X-many dollars.

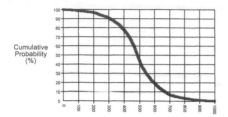

In this case, we can say that we have about a 42% chance of getting "at least" 500 or more.

FIGURE 14.12   Example of a "Z-shaped" cumulative frequency curve.

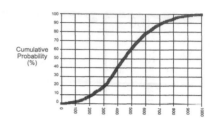

In this case, we can say that we have about a 40% chance of getting "up to 400 (including 0).

FIGURE 14.13   Example of an "S-shaped" cumulative frequency curve.

Most off-the-shelf risk-assessment packages offer the S curve as the default situation. However, those with experience in all types of risk assessment prefer the Z curve because its "high-probability" end is the end near the Y axis. This has great advantage when incorporating chance of failure into a risk assessment (see Chapter 16) or when calculating risk-weighted values (see Chapter 18), or when performing a host of other risk assessment tasks. Therefore, I recommend the Z curve as the most utilitarian projection of cumulative probability.

A typical misconception associated with cumulative frequency plots is that the "low-value" end of the curve (the end nearest the Y axis) should approach or reach zero. Consider the situation of two distributions, A and B, displayed as arrays of values in Figure 14.14. In each distribution, the smallest value is 2 and the value 2 only occurs one time in each distribution. The risk model that uses distributions A and B is simply using Monte Carlo random sampling to add A and B to produce C (C = A + B).

The value of least magnitude that C can attain is 4 (A = 2 and B = 2). Given that there are 20 values each in the A and B distributions, we have a 1 in 400 (20 × 20) chance of producing a value of 4. You will note, however, that the value of 4 plots at 100% on the Y axis in Figure 14.15. It should be apparent that there is not a 100% chance of producing a value of 4 by random selection from A and B (only 1 chance in 400). So, what do we have a 100% chance of producing with the equation C = A + B? There is a 100% chance of calculating a C-value of 4 *or greater.* Other X- and Y-axes pairs are similarly interpreted (a Y probability of getting X or greater than X).

This leads to another point concerning the interpretation of cumulative frequency curves. In Figure 14.15, the value of least magnitude on the plot is also the value of least magnitude that could have been produced from the distributions A and B. Similarly, the maximum value on the curve is 26, and that also is the value of maximum magnitude that could be calculated from distributions A and B (given our C = A + B equation). It is not typical that the end points of a cumulative frequency curve represent the minimum and maximum values that could have been produced.

Distributions A and B in Figure 14.14 each are comprised of only 20 values. If 400 or more Monte Carlo iterations are used to randomly sample distributions A

Variable A	Variable B
3	7
5	2
6	13
2	4
12	5
4	10
8	9
13	11
10	6
9	12
7	8
11	10
4	3
9	10
12	9
5	11
8	4
10	7
9	3
11	5

FIGURE 14.14   Two 2-value arrays for variables A and B.

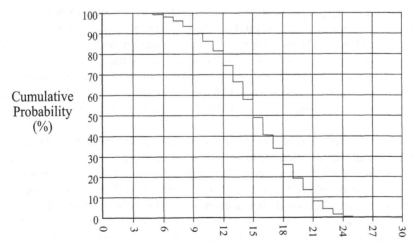

FIGURE 14.15   Cumulative frequency plot with a minimum value of 4 and a maximum value of 26. Curve created by summing variables A and B.

and B, then there would be a pretty good chance of producing the minimum and maximum values that could result from adding A and B (the greater the number of iterations, the better the chance of producing the minimum and maximum values). Distributions in actual risk-assessment models typically contain 100 to 500 or more values. Also typical is to have many more than two distributions in the risk assessment model. Consider a model composed of, say, two distributions, each containing 200 values. We would need to run tens of thousands of Monte Carlo iterations to have any reasonable chance of producing a cumulative frequency curve that had as its end points the values of largest and smallest magnitude that could have been produced by combination of the two distributions.

Because of the common complexity of risk assessment models, the speed of the typical computer, and other factors, most risk assessment models are allowed to run a number of Monte Carlo iterations that is between 1000 and 10,000 (the number of iterations generally can be set by the user and can be highly variable). This is far too few iterations to ensure that values of minimum or maximum magnitude have been generated. Therefore, the end points of most cumulative frequency curves do *not* represent the values of minimum and maximum magnitude that could have been produced. The end points represent the values of minimum and maximum magnitude that were produced in the given number of iterations.

At first blush, this would appear a weakness in the Monte Carlo process, but it is not. To illustrate the reason absolute minimum and maximum values should not be represented on the plot, consider the following situation. Suppose we are calculating profits from distributions representing number of units sold, price per unit, number of sales personnel, salary per hour for sales personnel, capital expenses, and operating expenses. If each of these distributions truly represents the range of values that could possibly occur (i.e., tails of the distributions are composed of values that are extreme and unlikely), there is a *very* small chance that, in the real world, there will be a simultaneous occurrence of the maximum number of units sold, the maximum price per unit, the minimum number of sales personnel, the minimum salary per hour for personnel, the minimum capital expense, and the minimum operating expense to produce the maximum possible profit. Although some may feel it would be useful to calculate the minimum- and maximum-possible profit, this can be done easily. A risk-assessment or Monte Carlo analysis is not needed to generate such values. The purpose of Monte Carlo simulation is to produce plots of things that are *likely* to happen. Alignment of absolute minimums and maximums, in the real world, is not likely. It also can be dangerous to produce such values.

Consider the situation of a company composed of several sales locations. The company would like to calculate the total combined profit from all stores. If each of the store-profit curves were allowed to range from the absolute minimum to the absolute maximum value ("biggest" and "smallest" values that could be produced), summation of these individual curves, even by Monte Carlo analysis (the way they should be summed) will generate "upside" and "downside" values that are highly unlikely and extreme. These values may imbue "management" with unrealistic expectations (these numbers tend to get recorded in accounting sheets and reported). These numbers also can be downright scary.

COMBINING MONTE CARLO OUTPUT CURVES

In the example above, the problem of combining (in this case, combining by summing) the Monte Carlo output profit curves from individual stores to arrive at a total-profit curve for all stores was considered. It is tempting (and frequently done) to add together all of the respective individual output curve minimums, means (or medians), and maximums. Another typical mistake is to sum (or otherwise combine) percentile values such as P10, P50, and P90 values.

Once again, this is invalid because it is highly unlikely that extreme values will, in the real world, align to produce a resultant (sum) value of minimum or maximum magnitude. To correctly combine Monte Carlo output curves, each output curve should be used as an input curve to a new Monte Carlo routine specifically designed to combine the curves. When this approach is implemented, real-world and reasonable combinations and ranges result.

7. COMPARING MONTE CARLO OUTPUT CURVES

15 Decisions and Distributions

CONTENTS

DECISIONS

The only time decisions have to be made is when something about a situation is unknown. Consider the scenario in which your spouse asks you to go to the store. You leave the house and arrive at the proverbial fork in the road. You look down the right fork and see no store; you look down the left fork and see the store. In this case, I submit, there is no decision to be made — the store is on the left fork and that is where you will go. However, if you reach the fork in the road and do not see a store down either fork, then a decision must be made. Even if you decide to stay where you are or to go back home, a decision still has to be made as to what to do. A decision must be made because something is unknown and there is uncertainty about something related to the problem.

Distributions are one means of expressing uncertainty in a problem and, in fact, using that uncertainty to our advantage. Therefore, distributions are used in decision making by affording a means of quantifying, expressing, and using uncertainty in the process of making a decision.

JUST WHAT IS A DISTRIBUTION?

One of the most common mechanisms for expressing uncertainty is the distribution. Just what does the term *distribution* mean? When you look at a distribution, just what should it be telling you?

Consider the situation of the new-hire engineer who has been summoned on his first day of employment into the office of the head of construction. The engineer is informed that as part of the plant-expansion construction, 2000 doors and door jambs must be built. The boss has been informed by the supplier of materials that the company will have to buy its lumber in bulk directly from the mill. This means that the lumber will not be custom cut, but rather will be available from the mill only in 1-foot increments; that is, the boards will come in 5-foot lengths or in 6-foot lengths or in 7-foot lengths, etc. The boss is unsure about what length lumber should be purchased, so she asks the engineer to measure the heights of several hundred adults in the community and present her with the measurement results so a lumber-length decision can be made.

Because it is the engineer's first day on the job, he is anxious to impress the boss. In his zeal, he visits the local university and arranges to borrow some laser equipment they use to measure distances. The engineer sets up the equipment to measure the heights of adults. The equipment is capable of measuring distances (heights) with great precision — to ten decimal places if needed.

After a week of collecting data, the engineer considers he has amassed a sufficient number of measurements and he produces a plot of the heights. A plot of a subset of the data appears in Figure 15.1. The measurements were carried out with such precision that there are no repeating values; that is, each value is unique. So, the frequency with which each value occurs is 1. The plot of the data that the engineer took to the boss is shown in Figure 15.2.

The boss takes one look at the plot and informs the engineer that the plot is not quite what she had in mind. She informs the engineer that she was expecting a more "typical" frequency plot with the bars of various height, and she sends him away to have another try.

The engineer wonders what is wrong with the plot he made. After all, in a data set of nonrepeating values, each value occurs once (the probability of any value

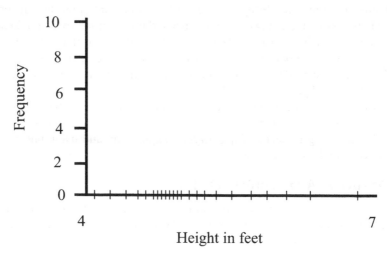

FIGURE 15.1 Measured values sorted and plotted on the X axis.

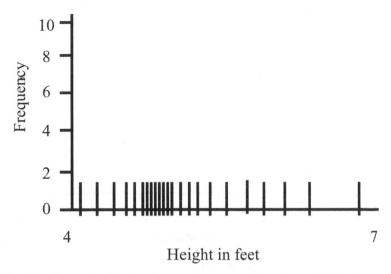

FIGURE 15.2 Plot of the X-axis values each represented by a bar of height 1.

occurring is one divided by the number of values in the data set). While this seemed perfectly reasonable to the engineer, he realizes that he would have to produce a plot that the boss would accept. To do this, he decides to change his way of thinking about the precision of the data he was presenting. Rather than consider each value individually, he comes to the conclusion that he will have to consider the measurements in "bins" or "buckets" or "groups." Figure 15.3 is the plot that results from considering the data in 6-in.-sized bins. The boss is happy with the plot of heights so the engineer too is happy, but something still bothers him about the "number of bins" thing. He decides, however, to forget about it for the time being because the boss has given him a new assignment.

The new task is to analyze some financial data. He is to get from the accountants data that show the range of profits the company, from preliminary estimates, might make from the plant expansion. The question to be answered is: How likely is it that the company will realize, in the first year of production, a profit of around $5M (the first-year economic hurdle for the project)?

After retrieving the data from the economists, the engineer loaded the data into statistical software and made three plots. These three plots are shown in Figure 15.4. When five bins were used to represent the data, the bar chart has a roughly lognormal appearance; that is, the plot is skewed to the right (remember, skewness direction is the direction in which the "tail" goes). When ten bars were used, the resulting plot indicates that there is a wide range of values of about the same frequency with less frequently occurring values on either side. When 40 bars were used, the plot takes on a bimodal appearance.

The changing appearance of the plot, however, is not the most disturbing aspect of the plots. The boss wants to know what the probability is of generating a profit of about $5M. According to the five-bar plot, the probability of making about $5M

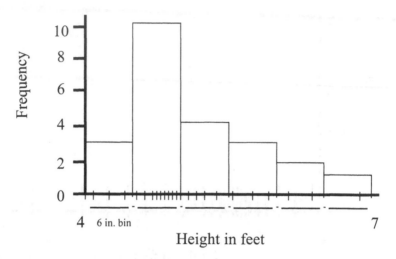

FIGURE 15.3 Plot of "binned" height data.

is about 50%. According to the ten-bar plot, the probability is about 30%, and the 40-bar plot indicates that the probability is no greater than 9%.

It is clear from the plots that this assignment must be more clearly defined. Just what did the boss mean by "about $5M"? If she meant the chance of achieving exactly $5M, and $5M just happened to be one of the X-axis values in the data set, then the probability of getting $5M is one divided by the number of values in the data set. If $5M is not one of the X-axis values plotted, then there is no chance of getting *exactly* $5M. However, the boss said *"about* $5M," so the engineer did not think she meant exactly $5M. Without clarification concerning the definition of *about*, he could not determine the level of precision (i.e., bar or bin width) with which he should represent the data. It should be clear from the three bar charts in Figure 15.4 that the probability of achieving "about $5M" depends on the bin size (bar width) which, in turn, is a function of the definition of *about*.

The lesson to be learned from this exercise is that at the contributing-factor-diagram construction stage, the group building the diagram needs to discuss and arrive at a consensus decision regarding the level of precision with which the data need to be considered.

So, to return to the original question posed at the beginning of this section, just what is a *distribution*? A distribution actually is the scattering of data points along the X axis as shown in Figure 15.1. A bar chart is a means of presenting the data, but it should be apparent that a presentation of the data should be commensurate with an understanding of the level of precision required.

For example, in our door building example, because the lumber is cut in 1-foot increments (5-ft long, 6-ft long, etc.), it does not matter to the boss whether the average height measured is 5 ft 5 in., or 5 ft 7 in., etc. She makes the simple deduction that she needs lumber at least 6-feet in length. However, if the engineer's charge had been to make space suits for NASA instead of doors for this company, this

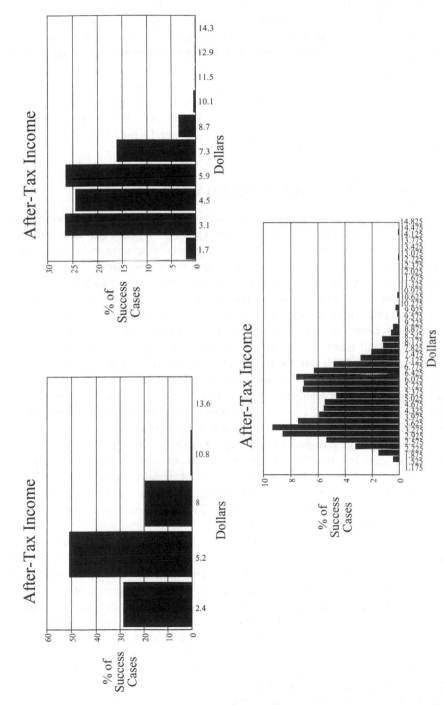

FIGURE 15.4 Frequency plots of common data using 5, 10, and 40 bins.

amount of imprecision in the data could not be tolerated (each suit is built specifically for each astronaut), and tolerances in height are very "tight."

A bar chart, then, is a representation of the distribution. The X axis of the bar chart represents the range of possible values for the variable being graphed. X-axis values denote the probability associated with a given bin size. The bin size should represent the level of precision required. When the contributing-factor diagram is being constructed, the group should discuss and come to consensus concerning the level of precision required and should agree that all plots made for the "answer" variable will respect this consensus. If this decision is not adhered to, very different representations of the data, such as those shown in Figure 15.4, can result. See the Output from Monte Carlo Analysis section of Chapter 14 for information on converting frequency plots to cumulative frequency space.

DISTRIBUTIONS — HOW TO APPROACH THEM

Those versed in risk assessment and/or statistics tend to classify distributions by their properties. Such names as "normal" (or "Gaussian"), "Poisson," and "gamma" are typical. Although this type of jargon is required if you want to be "a member of the club" and proliferate the erudite haze that shrouds statistics, I have seldom found such terms to be useful. Their limited use stems not so much from the fact that these names are relatively meaningless with respect to the actual distribution shape, but rather from the fact that my business has always been to bring risk assessment to people who are not necessarily versed in statistics or risk. Lawyers, accountants, plant managers, engineers, geoscientists, chemists, and others, while all capable in their fields, are likely not intimately familiar with the fundamentals or particulars of statistics or risk assessment. I always liken it to driving a car. I think I can drive pretty well, but I really do not know much about what goes on under the hood of my car. I count on someone qualified to take the complicated mess and make it easy for me to use, because I have to use the car.

That is my job with respect to statistics and risk assessment. Most people need to use these tools but, like the car, they really do not want to know what is going on under the hood, nor do they want to be asked to demonstrate a skill level or knowledge that is outside their ken. I like to say in my risk assessment classes that it is easy to make something hard, but it is really hard to make something easy. So it is with statistics and risk assessment.

I learned early in my career that if I were to be successful in bringing risk assessment to "the masses," then I would have to find a way to insulate and isolate others from the rigors and details of statistics. This required on my part not only an intimate knowledge of the subject matter, but also the development of algorithms, processes, and techniques that would allow those not steeped in statistics to use a risk assessment package and get the job done without having to provide information "unnatural" for them or to demonstrate extraordinary skills. I also learned early it is far better to use intuition, experience, and knowledge of the subject to build a distribution shape that best represents the data or variable than to represent a variable's data by algorithm (e.g., selecting a gamma or beta or chi-squared distribution) which is likely to be less useful. In most instances, users should be allowed to enter the data that they know about and should be able to shape the distribution to fit their mental model of the actual distribution of data. For example, an accountant may know the distribution of product prices for the next 6 months. Rather than being asked to provide an alpha-1 and alpha-2 value or some combination of mean and standard deviation, variance, kurtosis, or any number of other meaningless (to them) statistical values, the user should be able to input familiar data (price data, in this case). The user should be able to shape the distribution to fit historical price data, price projections, or experience-driven intuitive knowledge concerning the distribution shape. The algorithms I have developed to build distributions do just that without bothering users with a lot of statistical mumbo jumbo.

The first step toward "statistics-free" risk assessment was developing algorithms that would allow the generation and presentation of distributions by their general shape instead of by their statistical names. The names usually are tied to their properties (or inventors). To this end, algorithms were designed to allow the user to enter some combination of minimum, most likely, maximum, and peakedness values that would combine to create (mostly) tailed distributions. The names of the distributions denote the distribution shape. The primary distributions for which algorithms were developed are

- Symmetrical
- Skewed
- Spike
- Flat
- Truncated
- Discrete
- Bimodal

Reading data (and randomly sampling the data) and peakedness from a file will also be discussed in this chapter. I refer to the symmetrical, skewed, spiked, flat, and truncated distribution types as *simple* distributions, not because their mathematics or forms are simple, but because simple methods have been devised for constructing these distributions. The distribution-building algorithms generated can build any of these simple distribution shapes using select combinations of minimum, most likely, and maximum values. The combination of minimum, most likely, and maximum values needed to construct each distribution type will be conveyed in the following sections that describe each type.

SYMMETRICAL DISTRIBUTIONS

The normal (or Gaussian) distribution forms the basis for parametric (classic) statistical techniques such as the Student's t-test and others. Because this is a risk assessment book and not a treatise on statistics, and in keeping with my bent for practicality, the complex mathematics of such distributions will not be discussed here. In addition, in my quest to prod the reader to consider distribution form rather than function or equation, I will henceforth refer to such distributions mainly as *symmetrical*.

Some properties of symmetrical distributions include the following.

- They are bell shaped
- They are symmetrical about the mean value
- The mean and median values (and the mode, if there are repeating values as in the case of nominal data) are coincident

Figure 15.5 depicts a symmetrical distribution. This distribution type forms the basis for many classical statistical techniques. However, except in rare instances, I have found relatively little use in the world of risk assessment for the symmetrical distribution. This is not to disparage the statistical prowess and significance of the symmetrical distribution — it would be a faux pas of great magnitude to insinuate anything other than that it is the foundation upon which classical parametric statistical techniques are built. Having said that, I will now comment on the use of symmetrical distributions in practical risk assessment.

Experience with many individuals and groups has taught me a few things. Many times, a person performing a risk assessment must generate a distribution to represent a variable and does not have a credible notion as to the distribution shape that should be used. In such cases, there is a pervasive tendency for that person to turn to the symmetrical (normal, if you will) distribution type. I am not quite sure why this is so. Perhaps it is because this distribution plays such a significant role in classical statistics. Perhaps it is because we typically use the term *normal* to describe such distributions and the term somehow denotes to the person that it is "normal," so it is "normally" what should be used (imagine that).

When considering ranges around single values, it seems that phrases such as "plus or minus 10%" just roll easily off the tongues of those who really do not have an idea concerning the range or, more typically, simply have not given serious thought to the distribution shape. I only warn here that some thought should be given to the actual shape of a distribution before it is assumed that a "normal" distribution best represents a variable.

Figure 15.5 depicts a symmetrical distribution. Symmetrical distributions built by algorithms I have designed are tailed (tails of the distributions are asymptotic to the X axis) and are created by supplying a minimum and maximum value only (with the most likely value left blank). The concentration of values around the most likely value (i.e., the peakedness of the distribution) is discussed later in this chapter.

SKEWED DISTRIBUTIONS

The most prevalent distribution type is the skewed distribution. In my practical vernacular, a skewed distribution is defined as one in which the concentration of values in a frequency plot (the "hump," if you will) is not central. The lognormal distribution is probably the most popular example of the skewed distribution type. A skewed distribution is shown in Figure 15.6.

A collection of values representing those natural, business, or technical parameters will be skewed when sorted and plotted. This type of distribution can be skewed to the right or to the left (skewness is the direction of the tail). Figure 15.7

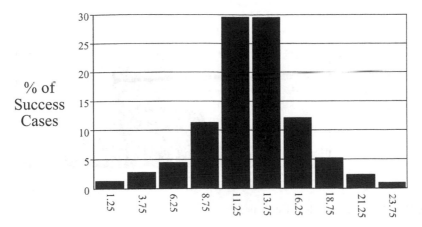

FIGURE 15.5   Symmetrical distribution.

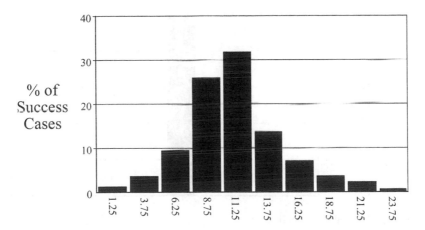

FIGURE 15.6   Skewed right distribution.

shows a skewed-left distribution. Skewed distributions built by algorithms I have designed are defined by entering a minimum, most likely (where the "hump" will be in the frequency plot), and maximum values. The tails of these distributions, as represented by a line connecting the tops of the bars in a frequency plot, are asymptotic to the X axis.

SPIKED DISTRIBUTION

In my terminology, a *spiked distribution* is defined as a distribution comprised of a single value. Representation of a parameter by a spike is often handy. Figure 15.8 shows such a distribution.

Depending upon the question being asked, a single value can be the optimal representation of the answer to that question. For example, if a risk assessment

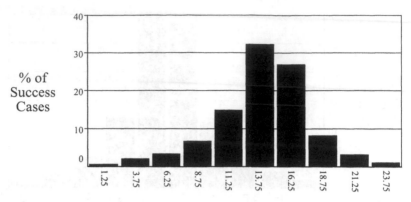

FIGURE 15.7   Skewed left distribution.

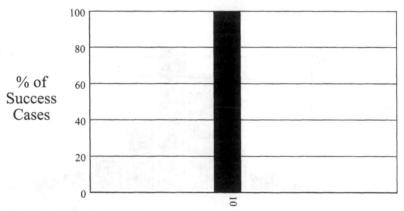

FIGURE 15.8   Spike distribution.

includes a parameter entitled "Number of people tested," this may in fact be a single value. The number of pages that make up this book is another example in which a single value, rather than a distribution of values, best answers the question.

Spiked distributions also are useful when attempting to determine whether or not a given risk assessment model is producing the expected results. In a risk assessment, when input parameters are represented by various types of multivalued distributions and are integrated using Monte Carlo analysis (see Chapter 14), the "answer" variables will themselves be distributions. With the resulting distributions being a "smear" of values, it often can be difficult to determine whether the risk assessment model is producing the desired result. In such cases, it is often instructive to represent each of the input variables, for the sake of model testing only, by a spike (single value). So doing will result in single-valued "answer" variables from which it is much easier to determine whether the model is producing the expected results. In the distribution-building algorithms I employ, supplying a most likely value only (leaving the minimum and maximum values blank) will produce a single-valued distribution.

FLAT DISTRIBUTIONS

What I refer to as the flat distribution goes by several other names in the books choked with technical jargon. In such volumes, these distributions are most commonly referred to as uniform distributions. Figure 15.9 shows a frequency plot that is the result of 1500 "random grabs" from a flat distribution.

Unlike the symmetrical or skewed distributions already discussed, the flat distribution has no concentration of values on the X axis. That is, the X-axis values of such a distribution are evenly or uniformly spaced, resulting in a frequency plot (bar chart or histogram) in which the heights of the bars are equal, that is, a flat (no hump) distribution. This implies that there is no most likely value.

The utility of flat distributions is apparent when the user has some idea as to the range of a variable but little or no idea what value is most likely. For example, in a given geographic sales area, an area manager may be in the process of opening a new store. For a given product to be sold in the store, the manager knows that the product sells for a minimum of $5.00 per unit (in an economically depressed part of her area) and for a maximum of $9.00 per unit in the "upscale" areas and in airport and hotel shops. The price for the item is set by what the market will bear. The new store is in a newly developed area for which there is no historical data, so the manager may have no idea what the ultimate price of the product will be. She does know, however, that the price will fall somewhere between the just-above-breakeven price of $5.00 and the maximum price of $9.00. In a marketing risk assessment model, the price for this item may best be represented by a flat (i.e., "don't know what the most likely value is") distribution.

The distribution-building algorithms I employ generate a flat distribution when minimum and maximum values are entered (leaving the most likely value blank as

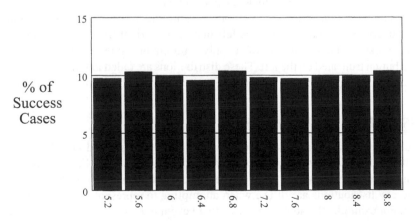

FIGURE 15.9   Flat (uniform) distribution.

with the construction of a symmetrical distribution) and when the user indicates the distribution is to be flat, rather than symmetrical.

TRUNCATED DISTRIBUTIONS

A truncated distribution is a single-tailed distribution. These distributions can be truncated on the right or left as shown in Figure 15.10. Truncated distributions are relevant when the minimum value or the maximum value also is the most likely value. For example, electric motors (furnace motors, pool filter motors, etc.) generally run at their maximum or governed number of revolutions per minute (RPMs). At times, such as when experiencing abnormally high resistance or when being turned on or off, the motor may run at less than the maximum RPM. However, running at the maximum RPM is the most likely condition. The RPM situation of such a motor would best be represented by a distribution that is truncated on the right — a distribution in which the maximum value also is the most likely value.

In the oil well drilling business, the restart time for offshore oil platforms (rigs) is a good example of a left-truncated distribution. Oil well drilling and producing activities may be halted on an offshore rig for a host of reasons. The approach of bad weather, a safety hazard, broken equipment, inspections, and a variety of other situations can cause a rig to halt operations. For a given rig in a given area there will be a minimum number of days required to restart operations on the rig regardless of how long or short a time operations were interrupted. Transport of inspectors to the rig, safety procedures, mandated start-up processes, and other required steps combine to prescribe a minimum restart time. For a given rig, the minimum restart time may be 2 days — there is no such thing as a restart time of less than 2 days. This restart time also is the typical restart time. That is, most of the instances of rig shutdown require the minimum number of days (two in this case) to restart the rig. However, occasionally operations on the rig are interrupted because of broken equipment that has to be replaced or damage to the rig by inclement weather. Restart times in these and other cases can range up to thirty days or more. Restart times, therefore, are best represented by a left-truncated distribution in which the minimum and most likely values are coincident (two days).

The distribution algorithms I employ require the user to supply a minimum and most likely value (maximum value left blank) for a right-truncated distribution and a most likely and maximum value only (minimum value left blank) to build a distribution truncated on the left. These distributions are tailed in that a line connecting the midpoints of the tops of bars in a frequency plot will be asymptotic to the X axis.

DISCRETE DISTRIBUTIONS

The discrete distribution is the first distribution type described here that cannot be built simply by supplying some combination of minimum, most likely, and maximum values (and clicking on a "flat" toggle button in the case of the flat distribution). Discrete distributions are of use when attempting to represent a discontinuous variable. For example, the admission price to an event may be $12.00 if you are between the ages of 5 and 65, $9.00 if you are a club member between the ages of 5 and 65,

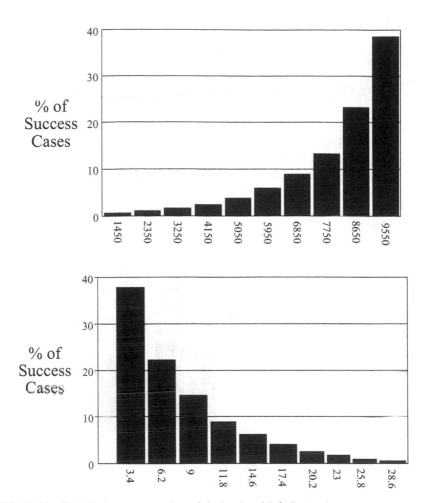

FIGURE 15.10 Distributions truncated on right (top) and left (bottom).

$5.00 if you are a full time student of any age, and $0.00 if you are over 65 years of age or are 5 years old or under. A risk assessment model may incorporate as an input variable the number of people in each price category who attended the event. Such a distribution is shown in Figure 15.11.

Admission price is a discrete, and not a continuous, distribution. In this distribution there are four discrete values: 0, 5, 9, and 12. When sampling this distribution in Monte Carlo analysis (see Chapter 14), there is no chance of drawing a value of, say, 6 from this set of values. Many real-world parameters used in risk assessment models are best represented by a discrete distribution.

The distribution-building algorithms I use will generate a discrete distribution when the user enters a distribution value (such as the 0, 5, 9, and 12 in our example) and a weight for each value. The weights entered are used to determine the relative frequency of the discrete values.

BIMODAL DISTRIBUTIONS

A bimodal distribution is simply one that contains more than one most likely value (i.e., more than one "hump" in a frequency plot or, in statistical jargon, more than one modal value). Such a distribution is shown in Figure 15.12. This bimodal distribution is typical in that it appears to be essentially two skewed distributions plotted in the same X–Y space. Such a distribution may be invoked to represent, for example, the return on an investment that depends upon the participation of other key investors. If we are to carry the investment alone, then the return may be in the lower range with a most likely return of around $23,500. However, if we are successful in attracting other investors, our return may be in the higher range with a most likely return of about $68,500.

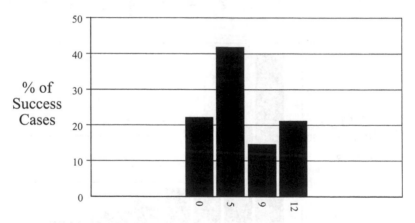

FIGURE 15.11   Discrete distribution.

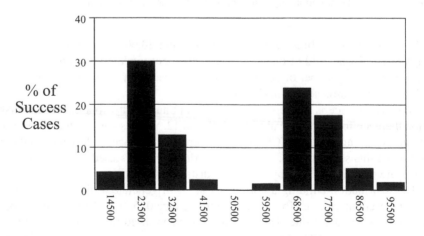

FIGURE 15.12   Bimodal distribution.

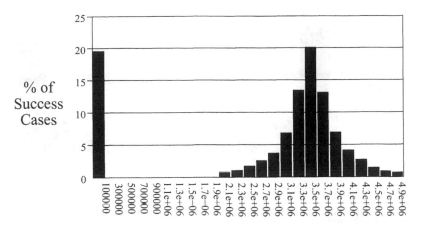

FIGURE 15.13 Bimodal distribution comprising spike and skewed distributions with associated weights of 0.2 and 0.8, respectively.

As previously indicated, a bimodal distribution is essentially two distributions plotted in a common X–Y space. The two distributions need not be skewed as shown in Figure 15.12. Nearly any combination of distribution shapes can be used to create a bimodal distribution. For example, we may have an investment that has the potential to return between $2 million and $5 million if the financial deal does not fall through. If the deal falls through, no return will be realized. There may be a 20% chance that the deal may fall apart, thus, a 20% chance of $0 return. In the bimodal distribution shown in Figure 15.13, there is a 20% chance of drawing (in Monte Carlo sampling — see Chapter 14) a 0 value and an 80% chance of drawing a value of between $2 million and $5 million. This distribution is a combination of spike and skewed distributions.

In yet another scenario, we may be attempting to represent with a bimodal distribution the amount of water processed by a water treatment plant on a given date. The plant may be in the process of being enlarged with the addition of new intake and output pipelines. If the construction is not finished on the date in question, then the plant will run at its old capacity. If, however, the construction of the new pipelines is complete or partially complete on the date in question, then the plant could process water at a rate somewhere between the old processing rate and the new one. However, the most likely scenario is that the plant will process water either at the old rate or at the new rate. In Figure 15.14, two truncated distributions (one truncated on the left, the other on the right) were used to create a bimodal plot (note that this plot is not unlike a beta distribution).

Most bimodal distribution-building technologies offered by risk assessment packages offer a means of emphasizing one of the distributions (modes) relative to the other. In the algorithms I use, a weight can be assigned to each of the two distributions that compose the bimodal. The weight values simply control the proportion of X-axis values assigned to each of the two distributions that are combined to produce the bimodal distribution. In the case of the distribution shown in Figure

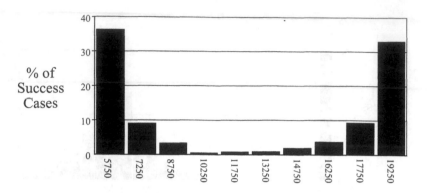

FIGURE 15.14   Simulation of a beta distribution by combining 2 truncated distributions in a bimodal distribution.

15.13, for example, a weight of .2 (20%) was assigned to the 0 (spike) value and a weight of .8 (80%) was assigned to the skewed distribution.

READING DATA FROM A FILE

Distributions are used to represent variables because rarely have we measured the entire parent population (i.e., we have rarely made all the measurements possible). Also, distributions are a convenient method of representing an entire population of values by requiring just a few input values from the person supplying the data (minimum, most likely, and maximum, or mean and standard deviation, or some such set of values).

At times, however, complete or nearly complete sets of data exist which may adequately represent the variable. In such cases, it may be tempting to feed this data directly into the risk model so that the Monte Carlo random-sampling engine can use the actual data as the basis of its sampling. There is nothing technically foul about sampling from "raw" data rather than from distributions, but there are a few potential problems that might arise.

Distributions built by software systems generally are composed of a discrete number of values. That is, between the minimum and maximum values of a distribution, there typically are 100 to 500 discrete values (this changes with both software package and distribution type) from which Monte Carlo analysis can randomly select. These values usually do not exist as a physical array of values, but are algorithmically generated at run time. Nonetheless, there typically are a limited number of values from which to sample.

One of the factors that dictates the number of Monte Carlo iterations required in a stochastic risk assessment model is the number of values that compose the distributions. The greater the number of values, the greater the number of iterations required to adequately sample the distributions (techniques such as Latin Hypercube can reduce the required number of iterations, but these techniques have problems of their own). The data values read from a file will be used as the basis for Monte Carlo sampling — just as in a distribution. When files containing large numbers of discrete values are employed, it should be obvious that a concomitantly large number of Monte Carlo iterations will be required to adequately sample the array of values in the file. This can have a significant impact on run time and risk-program complexity.

Having just warned about the foibles associated with files containing large numbers of values, I now will expound upon some of the problems associated with files that contain a relative dearth of values. It is a relatively rare occurrence that data collected over a finite length of time adequately represent reality. For example, an engineer on an offshore oil drilling platform may have recorded in a file the number of days the platform is shut down (downtime) each time there is a problem of great enough significance to halt operations. The engineer may have been record-ing data for 5 years. Typical downtimes have been in the range of 2 to 6 days per event. However, in the third year of downtime monitoring, a large "100-year" wave (a wave of this size is statistically expected to occur only once in 100 years) struck the platform, destroying critical equipment. The resulting downtime was 30 days. This value of 30 is one of only ten values that have been recorded over the 5-year period. Monte Carlo random selection will, on average, select the value of 30 one time out of ten. This, effectively, transforms the 100-year wave into the 10-year wave.

Except in the case of "short time" events, it is a rare situation that we have collected data over a sufficient length of time for our amassed data to represent the real world situation. Great caution should be exercised when substituting data files for distributions. Make every effort to ensure that a sufficient number of values has been collected so that the range and concentration of the values adequately aligns with the real world.

PEAKEDNESS

From the neverending battle to keep matters simple and practical, the notion of peakedness of a distribution was born. The typical user of a risk model is not versed in the concepts and nuances of statistical analysis including the concept of kurtosis. Kurtosis can be loosely (but not technically) defined as a measure of the peakedness of a distribution which is, in turn, a reflection of the concentration (density distri-bution) of values on the X axis of a frequency plot. Terms such as *platykurtic* and *leptokurtic* are used to describe distribution shapes. The kurtosis of a Gaussian (normal) distribution is 3.

Risk model users usually do not want to be bothered with this. What they would like to do is simply shape the distribution the way they wish without having to supply mean, standard deviation, kurtosis, or other statistical data. The concept of peakedness is a means by which they can do this.

The peakedness algorithm I employ (which can be replicated in any home-grown risk assessment package) is nothing more than an algorithm that controls the con-centration of X-axis values around the most likely value (the "hump") in those distributions which have most likely values. We explain *peakedness* to the user as being their confidence in the most likely value.

A peakedness range of 0 to 10 has been established. A slider bar is utilized to select a value between 0 and 10. A peakedness of 10 concentrates values near the

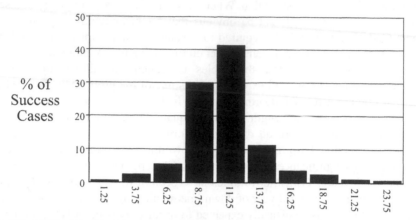

FIGURE 15.15   Distribution with a peakedness of 10.

most likely value, resulting in a frequency plot as shown in Figure 15.15. In this distribution, the probability of Monte Carlo random sampling selecting a value at or near the most likely value is relatively great compared to the probability of selecting a value in an equally sized range in the tail area of the distribution. By selecting a peakedness value of 10, the user is expressing relatively high confidence that values near the most likely value are those very likely to represent the variable.

A peakedness of 0 puts more values in the tails of the distribution at the expense of the most likely value. Figure 15.16 shows such a distribution. In a 0-peakedness distribution, values near the most likely are still more likely to be selected, but values in the tails of the distribution are relatively likely to be selected compared to a distribution with a peakedness value of greater magnitude. A low peakedness value indicates that the user has relatively low confidence in the most likely value. Using this peakedness scale, the user can express their confidence in their data and in so doing, control a major aspect of distribution shape.

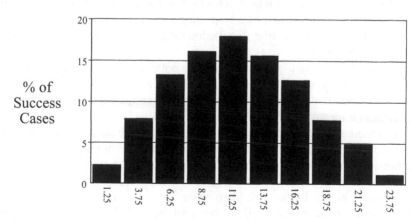

FIGURE 15.16   Distribution with a peakedness of 0.

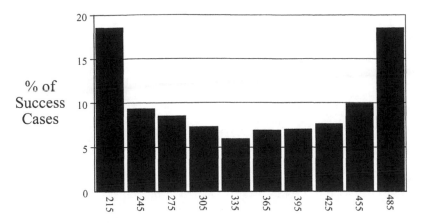

FIGURE 15.17 Beta distribution.

SPECIFIC DISTRIBUTION TYPES

For lack of a better term, I label as "specific distributions" all distributions that have been assigned names derived from their authors, their function, the algorithm that generated them, and other sources. Some of the most popular distributions of this category are the chi-squared, beta, gamma, Weibull, exponential, and error function.

A common thread of kinship with all of these distributions is that they require the user to supply distribution-building values that, typically, are difficult for the average person to provide or understand. The beta distribution (effectively 2 truncated distributions) in Figure 15.17 is an example. Construction of a beta distribution usually requires the user to provide alpha-1 and alpha-2 parameters. A gamma distribution will require you to provide alpha and beta values, a chi-squared distribution needs a nu value, and so it goes.

These distribution types can come in handy. For example, engineers may know from generations of experience that a given parameter is best represented by a chi-squared distribution. If this is so, then a chi-squared distribution should be used. However, in most situations, the average user will be at a loss to supply the statistical information required to build such distributions, and a simple distribution (as defined previously in this chapter) is a better choice.

16 Chance of Failure

CONTENTS

CHANCE OF FAILURE — WHAT IS IT?

There is no more common cause for spurious risk assessments and misinterpretation of risk-model results than the exclusion of chance of failure from the analysis. It will be illustrated in this chapter that consideration and integration of chance of failure, when appropriate, is crucial. *Chance of failure* (COF), as I write about and define it here, is the probability that the object of the risk assessment will experience abject failure. The total chance of success (TCOS) for the object of a risk assessment is calculated as follows.

$$\text{TCOS} = (1 - \text{COF variable 1}) * (1 - \text{COF variable 2}) * \ldots$$
$$(1 - \text{COF variable N}) \qquad (16.1)$$

where N is the number of variables that have an associated COF. Given this definition and calculation of TCOS, it should be clear that the application of the COF concept is valid only in cases in which failure of any variable that has an associated COF will cause the object of the risk assessment to fail. A physical analogy is the proverbial chain made of links that may be thought of, for example, as our project. A project risk assessment is composed of variables. If one link breaks, the chain fails. In the COF scheme, if one variable fails, the project fails. So, the concept of COF, as presented here, is *not* valid in situations in which failure of one component does not necessarily cause the entire enterprise to fail. A physical analogy is the strands of a rope. Failure of one strand (unlike failure of one link in the chain) does not necessarily cause the rope to fail.

A good real-world example in which COF might be applied is the launching of a new product in the marketplace. We may be considering, among other things, the price we need to get, legislation we need to have passed for our product, and the labor costs. If any one of these variables turns out, in the real world, to be outside our acceptable limits, then the product introduction will fail. That is, if we need a price of at least $12.00 per unit but all the market will bear is $10.00, then the game is over. Likewise, if the government fails to pass the required legislation, we fail. Similarly, we may experience labor costs above what we can bear, causing the enterprise to fail (no matter what other values we experience for other variables).

In contrast, some situations do not warrant the consideration of COF. For example, for our corporation to survive, we may need to attain an after-tax profit of $10MM. We arrive at our corporate profit by combining the profits from the 17 companies that comprise the corporation. If, in a given year, one or more of the companies "fails" (i.e., does not realize a profit), this does not necessarily spell failure for the corporation as long as the required profit can be gleaned from the companies that did show a profit.

A good real-world hybrid example is your car. Almost all components of your car have an associated COF. However, if we are building a risk assessment model to calculate our chance of the old car making a 1000-mile trip, consideration of failure for certain components is critical (transmission, alternator, radiator, etc.) while failure of other components (radio, electric locks, etc.) may be simply annoying.

FAILURE OF A RISK COMPONENT

Just how is it that failure is defined for a variable? The risk-model builder has to generate a definition for failure for each variable that has an associated COF. Let us begin with the situation in which COF affects the distribution.

Consider the simple case of the environmental engineer charged with supplying potable water in a drought-stricken area. Figure 16.1 shows two distributions: one for sediment load (percent of suspended solids in a liter of water) and one for size of the underground water reservoir (measured in cubic meters). The percent of rock particles in a liter of water in the given area has, according to historical data, ranged from a minimum of 3% to a maximum of 20% with a most likely sediment load of 10%. However, experience has shown that sediment loads greater than 14% are too high, causing the pipes to clog.

The percent of the histogram bars (frequency) in Figure 16.1 that will fall above 14% is the COF for sediment load. COF generally is expressed as a percent and is

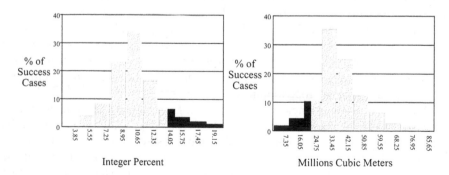

FIGURE 16.1 Frequency plots for sediment load (left) and reservoir size (right). Black areas represent % chance of failure (above 14% for sediment load and below 20mcm for reservoir size).

either calculated by the computer algorithm (as in the risk assessment package I use) or estimated by the user. The blackened area in the percent-load plot represents 10.5%, or a COF of 10.5. In this case, failure of the sediment-load variable is defined as the probability that the actual sediment load value will be greater than our maximum acceptable value of 14%.

Now let us consider reservoir size. The size of reservoirs in the area have ranged from a minimum of 3 million cubic meters (mcm) to a maximum of 90mcm with a most likely size of about 35mcm. Economic calculations indicate that if the reservoir is less than about 20mcm, then the reservoirs contain too little water to be economically developed. In Figure 16.1, the blackened area in the reservoir size histogram represents 5.5% of the area under the curve and, thus, a COF of 5.5. Failure for the reservoir-size variable is defined to be the probability that the actual reservoir size will be less than our acceptable minimum size of 20mcm.

In a risk assessment incorporating sediment load and reservoir size, the respective input distributions for the two variables would appear as those shown in Figure 16.2. Both of these distributions represent success-only values. Note that it is important not to allow the probability of the maximum-allowable sediment load or the minimum-acceptable reservoir size to go to near-0 probability.

The original sediment-load and reservoir-size distributions have essentially been "clipped" to produce the distributions shown in Figure 16.2. The parts of each distribution that have been clipped will be used in the calculation of the TCOS of the project.

$$\text{TCOS} = (1 - 0.15) * (1 - 0.055) = .85 * 0.945 = 0.803 \qquad (16.2)$$

CHANCE OF FAILURE THAT DOES NOT AFFECT AN INPUT DISTRIBUTION

In our water project example, one of the input variables is the size of the reservoir. Thus far, this variable has a single associated COF defined to be the chance that the reservoir is of a size that would yield uneconomic volumes of water. However, in the area where our water well is to be drilled, there is some chance that the desired

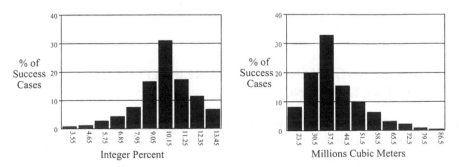

FIGURE 16.2 Sediment load (left) and reservoir size (right) distributions that have been "clipped" by their respective chances of failure.

reservoir rock type may have been altered by subsurface hydrothermal fluids (essentially, underground flowing hot water). The fluids may have transformed the reservoir to a rock type that is not of reservoir quality and, therefore, can yield no water at all.

It may be that we have measured the size of the potential reservoir rock just fine (using "remote sensing" techniques such as seismic or gravity-related technologies), but there is a chance that we are measuring the size of a "reservoir" of the wrong rock type. Therefore, we may define a second COF associated with reservoir size. This additional COF might be defined as the chance that we are measuring the size of the wrong thing (i.e., the "reservoir" is composed of a rock type from which we cannot extract water at all). This COF is associated with the reservoir size distribution, but does not clip or in any other way affect the distribution. For the sake of this example, let us say that we believe that there is a 10% chance that we have accurately measured the size of a body of rock of the wrong type.

Yet another variety of COF that does not influence input variable distributions is the type that is not associated with any input distribution at all. Continuing with our water project example, there may exist yet another reason for potential failure of the project. For our project to succeed, we need the local legislature to pass a law that exempts our project from regulations that preclude drilling in our proposed project area. Even though this project is being pursued at the behest of the local population, there is a significant chance that environmental groups may successfully block passage of the required legislation.

In this situation, we may define for our water project risk assessment model an input variable that is a COF-only (no associated distribution). The COF-only variable, in this case, might be defined as the chance that we will fail to get the required exemption. Because this variable has no distribution, it will not affect the Monte Carlo-calculated X-axis values but will affect the elevation of X-axis values to form our cumulative frequency plot and the projected Y-axis intercept of the curve. Let's assume that we believe there is a 20% chance that the required legislation will fail to be enacted. Given the two additional COF values we have added to our project, the TCOS for the project would now be:

$$TCOS = (1 - COF \text{ sediment load}) * (1 - COF \text{ reservoir size}) *$$
$$(1 - COF \text{ wrong rock type}) * (1 - COF \text{ legislation}) =$$
$$(1 - 0.15) * (1 - 0.055) * (1 - 0.1) * (1 - 0.2) =$$
$$0.85 * 0.945 * 0.9 * 0.8 = 0.578 \qquad (16.3)$$

INCORPORATING CHANCE OF FAILURE IN A PLOT OF CUMULATIVE FREQUENCY

In the preceding example, our total chance of success for the project is 57.8%. As described in the Output from Monte Carlo Analysis section of Chapter 14, the usual process for creating a cumulative frequency plot begins with sorting the X-axis values from smallest to largest. The Y axis is subdivided into a number of equally spaced segments (the number of segments equal to the number of Monte Carlo iterations) and the X-axis value of greatest magnitude is "elevated" one probability unit, the next largest value is "elevated" two probability units, and so on until the leftmost X-axis value is "elevated" N (number of iterations) probability units (Figure 16.3).

In the previous section of this chapter, we calculated the TCOS of the water project to be 57.8%. To incorporate COF in the cumulative frequency plot, rather than subdividing the 0 to 100 Y-axis range into N (number of iterations) segments, the 0 to TCOS (in this case, 0 to 57.8) range is subdivided into N equally spaced segments. The X-axis values are "elevated" as described previously, resulting in a cumulative frequency plot whose projected left end would intersect the Y axis at the TCOS value (Figure 16.4).

Figure 16.5 shows the same water project data plotted without considering COF. Note that the probability of achieving a dollar benefit amount of $90,000 or greater is about 86%. When considering COF, as shown in Figure 16.4, the probability of

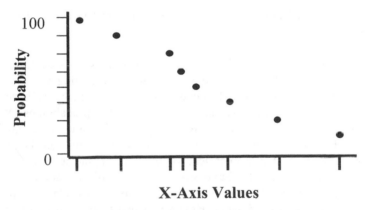

FIGURE 16.3 Individual X-axis values raised parallel to the Y axis according to their cumulative probability.

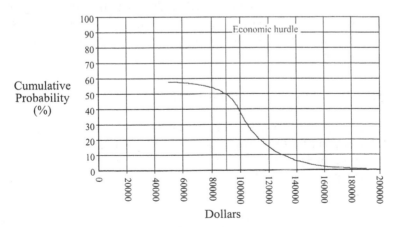

FIGURE 16.4 Cumulative frequency plot that includes chance of failure.

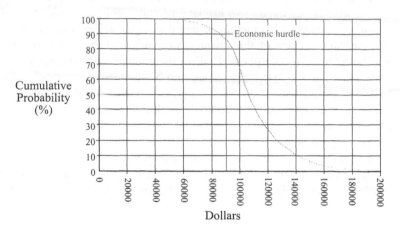

FIGURE 16.5 Cumulative frequency plot equivalent to that in Figure 16.4 but without chance of failure.

achieving the same dollar benefit is only about 50%. This is a significant difference that should result in different decisions being made concerning the water project.

It should be noted that when the cumulative frequency curve is allowed to "rise" to 100% on the Y axis (no consideration of COF), the probabilities (Y-axis) and percentiles (X-axis) are correlated. That is, the X-axis value, for example, which represents the 50th percentile (i.e., 50% of the X-axis values are larger than this value and 50% are smaller) also projects on the Y axis at the 50% probability point. Likewise, the 90th percentile X-axis value (i.e., 10% of the values are larger than this value) projects at 10% probability on the Y axis. This relationship holds true for all percentile/probability pairs. However, when COF is incorporated into the cumulative frequency plot, this percentile/probability relationship no longer applies.

Incorporation of COF does not affect percentiles. For example, the X-axis value that represented the 50th percentile prior to incorporating COF still is the 50th percentile value after COF is considered (COF does not affect the arrangement of X-axis values). However, when the cumulative frequency plot is "depressed" by COF, the *probability* of obtaining the 50th percentile point (or a value greater than it) no longer is 50%.

ANOTHER REASON FOR CHANCE OF FAILURE

In the water project example, the cumulative frequency curve was appropriately depressed to account for COF. This is all well and good; however, let us consider the consequence of ignoring COF in a simple risk model in which COF does influence the input variable distributions.

Consider the simple model of

$$\text{Income} = \$/\text{Package} * \text{Shipment Rate/Day} \qquad (16.4)$$

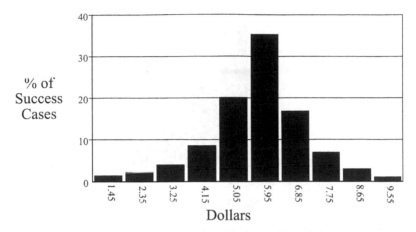

FIGURE 16.6 Frequency plot of dollars per package.

in which $/Package and Shipment Rate/Day are distributions with associated COF values. The $/Package distribution may range from $1.00 to $10.00 with a most likely value of $6.00. The associated COF will be defined to be the chance that we will realize a $/Package rate less than the required $4.00. Below this rate, a partner company that shares in the $/Package rate will pull out of the enterprise and, consequently, our business will fail. This distribution appears in Figure 16.6.

The independent Package-Shipment-Rate-per-day (PSRPD) variable is represented by a distribution with a minimum of 2000, a maximum of 20,000, and a most likely of 8000. Our business could survive with a minimum PSRPD of 2000; however, if the PSRPD gets much greater than 14,000, a significant bottleneck is created at the warehouse which causes our business to shut down for an unacceptable length of time. So, the associated COF will be defined as the chance that the PSRPD will exceed our warehouse capacity. This distribution appears in Figure 16.7.

If we use $/Package and PSRPD distributions as inputs to a Monte Carlo–based risk model that calculates income from the (admittedly oversimplified) equation

$$\text{Income} = \$/\text{Package} * \text{PSRPD} \qquad (16.5)$$

then we could, on these Monte Carlo iterations, draw the following $/Package and PSRPD values:

$/Package	*	PSRPD	=	Income
3		10,000		$30,000
5		6000		$30,000
2		15,000		$30,000

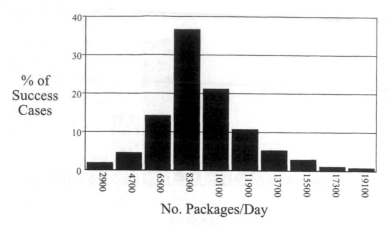

FIGURE 16.7 Frequency plot of package shipment rate per day.

Each of the random selections draws $/Package and PSRPD values which result in an income of $30,000. However, two of the three iterations include values that fall within the failure range for one of the variables. Therefore, given our three-iteration case, two thirds of the values that would plot at $30,000 on a histogram of income would not have succeeded in the real world — the histogram would significantly exaggerate our chance of achieving $30,000 (in this instance, our chance of achieving $30,000 is only one third of the histogram's indicated probability).

It is tempting to suggest as a solution to this problem that we simply "clip off" the failure parts of the input distributions, but that is tantamount to hiding our heads in the statistical sand. The values that populate the failure sections of distributions such as these do occur in the real world and must be reckoned with. When the concept of COF is employed, the distributions are "clipped" to exclude the failure values (sometimes the failure area can occur in the middle of the distribution), but are accounted for by "depressing" the cumulative frequency curve so that probabilities interpreted from the curve reflect the COF.

THE "INSERTING 0s WORK AROUND"

To nearly everyone who comes to grips with the concept of COF (especially spreadsheet bigots, it seems) it eventually occurs that: "Hey, I could just calculate a 0 when I run into a "failure" situation and I won't have to fool around with this COF stuff." I can tell you from experience that the route of "substituting 0s for failure" is the road to ruin.

For example, it is tempting to consider inserting some logic in the risk model that will recognize a failure value and so calculate a zero. In our $/Package and PSRPD example you might write

$$\text{If } \$/\text{Package} < 4 \text{ or PSRPD} > 14 \text{ then Income} = 0 \qquad (16.6)$$

and think you have taken care of that nasty failure problem. Well, there are several problems with this approach.

First, some risk assessment models are quite complex and can consume a significant amount of time and resources for each iteration. Typically, it takes no less time to generate a zero result than to generate a nonzero result. If the situation includes a relatively high COF, most of the time and effort is spent generating noncontributing zeros.

In this same vein, if a 1000-iteration risk model includes a 50% COF, this means that 500 zeros will be generated and that the cumulative frequency curve will be composed of only 500 "nonfailure" values. This would not be so terrible if each situation you needed to run through the model had exactly the same COF. It is the usual situation, however, that individual projects have uniquely associated COFs. Using the 0-insertion method, you will be left with a curve comprised of 500 points (50% COF) from one run of the model, a second curve made up of 300 points (70% COF) from a second run of the model, and so on. Not only are the curves composed of relatively few values, but the disparity in values can make statistical comparison of curves difficult or invalid.

The clever person might look at this disparity of non-failure values problem and conclude: "Hey, I know what I'll do, I'll just let the model run until it generates, say, 1000 nonzero values." Nice thought. More iterations generate more zeros. More iterations also increase the chance of sampling and combining extreme values in the tails of the distributions. Under this more-iterations scheme, two output cumulative frequency curves which are generated by combining the same input distributions but have different TCOS values can have significantly different X-axis ranges. This is because more iterations tend to extend the end points of the cumulative frequency curve. This can make difficult comparison of curves from multiple runs of the risk model. A cumulative frequency curve generated using the COF concept will always contain the same number of points as the number of Monte Carlo iterations.

As is my wont, I have saved the best for last. The most glaring problem associated with "substitute zeros" is that sometimes zero is a nice benign value that can be used as a failure indicator and sometimes it is not.

Consider the typical risk model. There usually exists "up front" a technical section followed by, at least, a section that calculates economics (sometimes an engineering section might be sandwiched between the technical and economic parts). In any event, zero might be able to represent failure in one part of a model but would not be interpreted as failure in other sections. For example, it might be fine to calculate factory output in the first part of a risk model. Factory output is either something (a number greater than zero) or nothing. There likely is no such thing as negative factory output. So, it might be tempting in this case to represent with zero the failure of the factory to produce. An output curve of production, then, might include many zeros indicating the number of times the Monte Carlo process calculated no production.

In a multistep risk model, however, the output from one step is passed along to the next. It may well be that zero can represent failure in the calculate production step, but a subsequent engineering step (in which there can be successful and positive values) or economics step (in which deficit spending does not represent failure) may not treat zeros as the benign values they were in the production stage of the model. Passing along strings of zeros from one part of a risk model to another is a dangerous practice.

COF AND MULTIPLE OUTPUT VARIABLES

A cautionary note concerning the application of the COF concept with multiple output variables. Consider the case in which we are computing two output variables A and B from input variables C, D, E, and F. The equations might be

$$A = C + D \text{ and } B = E * F \tag{16.7}$$

Each of the four input variables may have an associated COF. In this case the risk model should calculate two TCOS values; one for A from the COF values associated with C and D, and one for B from the COF values for E and F. Given the separate and independent equations for A and B, the failure of, for example, E has no affect on A. Likewise, the failure of either C or D will not be considered in calculating the TCOS for B. Great caution should be taken when constructing risk models so that TCOS values for individual output variables are correctly calculated.

17 Time-Series Analysis and Dependence

CONTENTS

INTRODUCTION TO TIME-SERIES ANALYSIS AND DEPENDENCE

Time-series analysis and dependence are functions that help risk models better emulate actual situations. Time-series analysis allows us to break free of the single-period assessment and to project the analysis through time. This is a facility that is critical in most business applications, for example, in which cash flows over time are combined to generate project-representative values such as net present value (NPV). Dependence is a technique that allows a risk model to respect the relationships between variables. Without application of dependence technology, any hope of emulating real-world processes is folly. Both time-series analysis and dependence are treated in the following sections.

TIME-SERIES ANALYSIS — WHY?

The desired outcome of a risk analysis is, of course, a stochastic assessment of the situation at hand. In a business setting, regardless of the technical, environmental, political, or other nature of the risk model, ultimate decisions are based primarily on a probabilistic financial assessment of the opportunity. Financial measures as discounted return on investment (DROI), internal rate of return (IRR), or NPV are typical factors upon which decisions are based. Financial considerations of this ilk require the evaluation of an opportunity over time, commonly some number of years. An NPV, for example, is based upon a series of cash flows.

It is beyond the scope of this treatise to explain the nuances of the calculation of measures such as NPV, the definition of which can be found in nearly any basic financial text. It is, however, precisely the premise of this chapter to explain how a risk model can result in the calculation of yearly cash flows and financial measures such as NPV.

TIME-SERIES ANALYSIS — HOW?

Economists working for most companies these days typically employ spreadsheets of one type or another to handle the financial attributes of a project. It is quite common to see in such spreadsheets some number of columns designated as years with rows below the yearly column headings representing variables in an analysis. Variables such as yearly labor costs, capital costs, yearly production, and other parameters are typical (see Figure 17.1). In a deterministic spreadsheet, these parameters and others would be combined to calculate, for example, a single NPV. In the risk assessment world, however, we strive for a distribution of answer-variable coefficients.

Year	1998	1999	2000	2001	2002
Labor costs (in thousands $)	900	600	550	500	500
Capital costs (in thousands $)	1000	500	400	400	400
Yearly production (tons)	30000	35000	40000	45000	45000

FIGURE 17.1 Spreadsheet with time periods as columns and variables as rows.

Generating a distribution for an answer variable such as NPV is accomplished by combining input variables that are themselves distributions. Using spreadsheets and a spreadsheet add-on risk assessment package, the yearly and single-valued parameters (i.e., labor costs, etc.) can have each of their yearly values expanded into a distribution. That is, rather than having the 1998 labor costs in Figure 17.1 represented by the single value of 900, a 1998 labor costs range might be from a minimum of 700, a most likely of 900, and a maximum of 1000.

Several techniques may be employed to transform single values into distributions. For example, it may be specifically known that 1998 labor costs will range from 700 to 1000 and that 1999 labor costs will range from 500 to 750 and so on for subsequent years. In the risk assessment software, then, a separate distribution can be established for each year for each variable. This technique has the advantage of allowing the risk model user to specifically control the parameter values through time. Several disadvantages are inherent in this methodology. First, it is rare, indeed, that we in fact know specific ranges for years beyond the current year. Also, it should be clear that implementing this type of single value expansion puts the onus on the user to build an inordinate number of distributions (one per year per variable). Generally, this is impractical.

Another method commonly employed to expand single values into distributions is to establish one or more expansion distributions. For example, we may believe that our labor costs in the first 3 years of the project will range from 10% less to

20% greater than the single value in the spreadsheet. We can then, for the first 3 years, multiply the single spreadsheet value by a distribution that ranges from 0.9 to 1.2. If we believe that for some number of subsequent years different ranges are valid, other expansion distributions can be established and applied in the appropriate time frames. This method of single-value expansion reduces the year-to-year control of the parameter distribution, but also reduces the number of distributions the user is required to supply. On each iteration of the Monte Carlo process, the 0.9 to 1.2 distribution will be sampled and the randomly selected value will be multiplied by the single value in the spreadsheet.

Thus far, the time-series examples in this chapter have involved the combination of a risk modeling package and a spreadsheet. However, any complete and self-respecting risk assessment software package should have the self-contained where-withal to effect time-series analysis without need of a spreadsheet. Such risk packages must contain a means by which any number of time series can be established and a syntax for manipulating multivariate analyses over time.

TIME-SERIES ANALYSIS — RESULTS

A risk analysis equation generally terminates in the calculation of a resultant variable that is represented by a distribution. Because a time-series analysis generally considers a proposition over a number of years (i.e., a distribution of years), a resulting time-series, answer-variable plot must be able to depict a distribution of distributions. Such a plot is shown in Figure 17.2.

In Figure 17.2, each vertical bar represents the distribution of values for 1 year in a time series. The short horizontal bars at the bottom and top of each vertical bar represent the minimum and maximum values respectively for that time period. The third horizontal bar between them represents either the mean or median value of the distribution. Positioning the computer's cursor on any vertical bar and "clicking" on

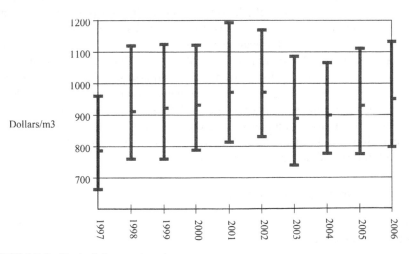

FIGURE 17.2 Typical time-series plot.

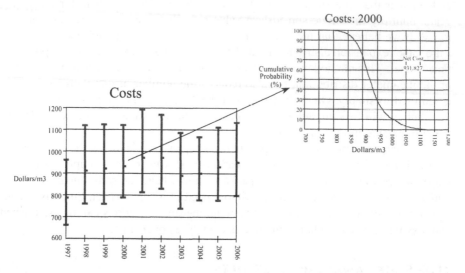

FIGURE 17.3 Time-series plot with "expanded" single period.

it can result in the display of the distribution for that time period (Figure 17.3). In this example, a time series of cash flows was generated. From such a time series, an NPV distribution for the project can be calculated.

In reality, a time series plot such as that in Figure 17.2 "knows" nothing about time. X-axis increments can represent any set of time periods such as months, days, or hours. In fact, the X-axis intervals need not represent time at all. For example, each tick mark on the X axis might represent a different investment opportunity with each vertical bar indicating the range of return for that investment.

A conventional time-series plot is just that — a plot of a series of time periods. The bars on the plot represent the range of answer-variable values for a given time period. A time-series plot itself, however, actually has no predisposition toward delineating time on the X axis. That is, the tick marks along the X axis need not represent segmented time. For example, consider the corporation with seven operating business units. Corporate management might be interested in comparing the net incomes from each unit for a given year. A time-series plot can be constructed in which the X-axis tick marks represent the names of business groups while the Y axis, in this case, represents net income in dollars. Such a plot appears in Figure 17.4. Similarly, a law firm might wish to compare strategies concerning a potential litigation situation. In this case, X-axis tick marks could represent options such as settle, arbitration, adjudication, or jury trial.

SOME THINGS TO CONSIDER

Time-series analysis is a powerful analytical technique and is an often-employed process in Monte Carlo–based risk models. The marriage, however, can result in some algorithmic behaviors of which you should be aware. In traditional Monte

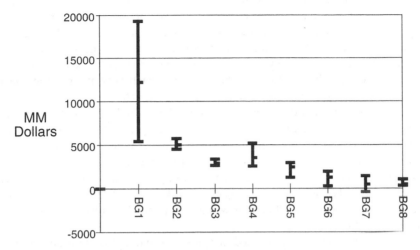

FIGURE 17.4   Business group net income time-series plot in which X-axis entities are not time periods.

Carlo analysis, each distribution representing a risk-model variable is sampled once on each iteration. Classic Monte Carlo analysis would, on each iteration, combine the selected values according to the equations to arrive at an iteration-specific answer. The introduction of time-series analysis complicates this simple scheme. If, for example, we decide to calculate NPV based on a 5-year projection of cash flows, then on each iteration the set of equations would be executed five times — once for each year.

On the face of it, this seems reasonable and interjects only a slight wrinkle in the fabric of traditional Monte Carlo processes. However, consider the following situation. We have projected margins for the next 5 years. We expect our margin to grow over the years by 10 to 20%. This growth is represented by a growth distribution that ranges from 0.1 to 0.2. Multiplication of the year's margin value by a randomly selected value from the growth distribution will cause our margin to increase by between 10 and 20% — just what we wanted. Or is it?

Recall that each distribution is sampled once on each iteration. Remember also that on each iteration when we employ time-series analysis, we apply the randomly selected values in each period of the time series — in this case, in each of 5 years. So, if on iteration 1 we select a growth rate of 15%, we will grow our margins at 15% every year. Is this what was intended and is this what happens in the real world?

Real-world yearly growth is somewhat erratic. That is, margins may grow by 12% this year, by 17% next year, by 10% the following year, and so on. If we want our risk model to emulate the real world, then we need to employ a mechanism that allows the sampling of distributions not only by iteration, but by time-series period. The risk assessment software I employ facilitates this type of period-based sampling and analysis. You should also select a software package that accommodates this process.

Well, so what if I cannot or do not resample the growth distribution on each iteration? What possible difference could it make? Turns out, a big one.

Consider the situation in which we do not resample the growth distribution across periods. In a many-iteration model (it is not uncommon to run such models for thousands of iterations), it is statistically safe to say that in a distribution that ranges from 0.1 to 0.2, on at least one iteration a value very near 0.2 will likely be selected. Likewise, it is likely that a value at or near 0.1 will be selected on another iteration. Because these values will be applied in each year, the resulting year-by-year plot of margins will have a much greater range (for each year) than would a plot resulting from period-by-period sampling of growth. In the single-selection case in which 0.2 was selected as the growth rate, each year will grow by this amount. This will generate a significantly higher margin upper bound than if the growth rate were resampled for each period. The value of 0.2 might be selected for year 1, but lesser values are likely to be selected for subsequent years. This will significantly lower the upper bound for margin projections. Likewise, selection of 0.1 and employment of that value across all periods will give an unrealistically pessimistic view of margin growth.

So now, having been convinced that this is a real and acute problem, we have purchased or built a risk system that can accommodate period-based sampling of distributions. So now we have it licked, right? Wrong. There is a little item called dependence that has to be considered.

Traditionally, when dependence between two variables is established, one variable is designated the independent variable and the other the dependent variable. This Monte Carlo process first randomly samples the independent variable. Then, depending on the degree of correlation between the two variables, the process establishes a dependent-variable value that is reasonable for the independent-variable coefficient. This process happens once for each iteration.

Now we are resampling our distributions not only for each iteration, but across N periods within an iteration. If the distribution being resampled period by period has dependence relationships with the other variables, then those relationships need to be honored across periods, not just across iterations. Again, a risk-model-building software package should be purchased or built that will accommodate such a process.

DEPENDENCE — WHAT IS IT?

In the real world, parameters that are co-contributors to a risk assessment may be independent. For example, as part of a Total income calculation a risk model might contain two distributions for the variables Number of states in which we will operate in the year 2000 and Number of countries in which we will operate in the year 2000. In this business, the domestic and international arms of the corporation are separately managed and funded. In the risk model, then, the aforementioned distributions can be independently sampled and no invalid pairs of values can be generated (i.e., in the risk model, any number of states can be combined with any number of foreign countries).

In an actual risk model, many variables exhibit dependent relationships. For example, in the risk model we use in the domestic part of the corporation, our income calculation considers variables such as Stage of technology and Customer commitment. For both variables, we might generate a distribution. In the risk model, we would not want to independently sample these two distributions because such sampling might combine, in a single Monte Carlo iteration, high Stage of technology values with unrealistically low values for Customer commitment. These two variables have a relationship in that as our corporation is increasingly viewed as the leader in our technical field, the commitment of customers (percent of customer base, number of long-term and binding contracts, etc.) increases. The concept of dependence is that of honoring the relationship between two or more variables.

INDEPENDENT AND DEPENDENT VARIABLES

To invoke the concept and underlying technology of dependence, the first item on the agenda is to decide, for a pair of variables exhibiting dependence, which is the independent and which is the dependent variable. That is, which variable controls the other?

In the case of the Stage of technology and Customer commitment variables, it may be clear in our corporation that our customer commitment increases as a result of our technical prowess (as opposed to our technical ability increasing because we are exposed to more and varied customers). In a different company, however, increased technical ability might result from the diversity and size of the customer base.

In our situation, Stage of technology would be considered the independent variable and its distribution randomly sampled by the Monte Carlo process. The value selected from the Customer commitment distribution would, on each iteration, depend upon the Stage of technology value selected. Thus, Customer commitment is the dependent variable.

For cases in which it cannot be easily discerned which is the independent (controlling) and which the dependent (controlled) variable, the independent variable should be selected based upon your confidence in a given variable. For example, in another company, it may not be clear which of our two variables (Stage of technology or Customer commitment) is the controlling entity. They may know, however, that the values in our Customer commitment distribution were gleaned from a comprehensive, formatted, and rigorous survey of all customers. Values that comprise the distribution for Stage of technology were generated in a meeting of managers, many of whom were are not intimately familiar with the technology. In this case, they will have much more confidence that the Customer commitment values actually represent reality and, therefore, the Customer commitment variable would be selected as the independent variable.

DEGREE OF DEPENDENCE

When it has been established in a risk assessment that we indeed have a dependent relationship between two or more variables, and when we have decided which of the variables will serve as the independent variable and which as the dependent, the next hurdle is to establish the strength of the relationship (dependence) between the variables. The strength of a relationship can be calculated or estimated in several ways.

The strength of a dependence usually is expressed as the degree of correlation between two variables (see Selected Readings for this chapter for references relating to correlation). Correlation can loosely be described as the degree to which one variable "tracks" another. Correlation is usually expressed on a scale from −1 to +1 with 0 indicating no correlation at all (no relationship between the variables).

Figure 17.5 is a graphical representation of two variables that are independent; i.e., they are uncorrelated and have a correlation coefficient of 0. Two such variables might be the number of eggs used on a day of business in the corporate cafeteria and the number of cars that passed by the building that day. Because the number of eggs used and the number of cars passing by have nothing to do with one another, for any number of cars (any X-axis value), any number of eggs (any Y-axis value) can be chosen. So, knowing how many cars passed by on a given day does not help at all in predicting how many eggs were used that same day. There is no correlation and, therefore, a correlation coefficient of 0 is appropriate.

The opposite end of the correlation spectrum is perfect correlation indicated by a correlation coefficient of −1 or +1. A value of −1 indicates perfect negative correlation and +1 indicates a perfect positive relationship. Well, what does that mean?

The minus (−) part of the −1 correlation coefficient means that as the value of one variable increases, the value of the other variable decreases. The +1 means that the correlation is without variance. A good real-world example is the gumball machine in the company lounge. Concerning the gumball machine, we may record how many gumballs are in the machine each day (assuming it is not refilled) and how many coins are in the machine. After recording the number of gumballs and the number of coins each day for 6 months, we plot up the data and find that it looks like the plot in Figure 17.6.

From this plot, we find that if we know the number of coins, we can predict exactly (correlation of 1) the number of gumballs left. We also know that this is a negative relationship because as the number of coins goes up, the number of gumballs goes down.

Figure 17.7 shows a plot of perfect (1) positive (+) correlation. This is a plot from Corporate Security which indicates what they wish were true. The plot shows the difference between the number of access-card readings from people entering the building minus the number of cards read when people left the building plotted against the number of people remaining in the building. According to Security theory, the

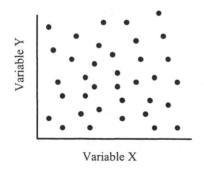

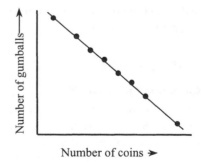

FIGURE 17.5 Cross plot of 2 independent (uncorrelated) variables.

FIGURE 17.6 Cross plot of 2 variables with a correlation coefficient of near −1.

number of cards read for people coming in less the number of cards read for people exiting should exactly match the number of individuals in the building at any given time. So, knowing the difference in card readings (X-axis values) should exactly predict the number of people (Y-axis values). This should be a perfect (1) positive (as values for one variable increase, values for the other variable increases) relationship.

The fire marshal has visited us and has noted that we have only two fire doors. She believes that we have more people in the building than only two doors can handle and, thus, thinks we need at least a third door. Corporate Security has been charged with counting noses.

Security thought that it could tell exactly how many people were in the building by examining their card-reading data. However, several counts of bodies revealed that the actual number of people in the building was not exactly predicted by the card reading data. Security discovered that it is typical, on both entering and leaving the building, that one person will open the door and several people will enter or exit while the door is open.

Therefore, while it is true that the number of card readings is some indication of the number of people in the building, it is not the perfect predictor (+1) they would like. So, if the number of cards read does give them *some* indication of body count, the correlation between the two variables is not 0. Likewise, it is not +1 because it is not a perfect predictor. It follows, then, that the relationship between these two variables should be indicated by a value somewhere between 0 and +1.

If historical data are available (as in the case where Corporate Security has card reading and actual body count data), the typical method employed to establish a relationship strength is to load the data into a software package that will perform a linear (or nonlinear, if you want to get really fancy) regression. Linear regression models will return an r value (correlation coefficient) that will give a reasonable estimate (for linearly varying data) of the strength of the correlation or relationship between the two variables. That is, how good a predictor of the dependent variable (body count) is the independent variable (card readings)?

Again, 0 indicates that it is no good at all; 1 indicates that it is perfect. Neither of these things is true in this case. Therefore, a plot of card readings vs. number of bodies would look something like the plot in Figure 17.8. A regression model run on these data might return a correlation coefficient of, say, 0.8. This indicates that for any value of card readings selected, there is some reasonable range (Y1 to Y2 in Figure 17.8) of Y-axis values (number of people in the building).

Remember, a correlation of 1 yields a single point on the Y axis for any X-axis value selected. A correlation of 0 means that for any X-axis value selected, any Y-axis value is valid. So as we move from 1 to 0, the range of Y-axis values that would be reasonable "matches" for a given X-axis value gets greater and greater until, at 0, the entire range of Y-axis values is available.

So if we determine that there is a dependence between two variables, and if we then decide which variable is independent and which dependent, and if we assign, based on historical data and regression analysis, a correlation coefficient to the relationship, then the dependent variable will track the independent variable. That is, on each iteration of Monte Carlo analysis, the independent variable will be randomly

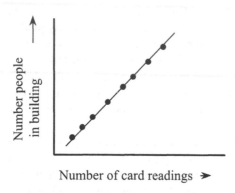

FIGURE 17.7 Cross plot of 2 variables with a correlation coefficient of near +1.

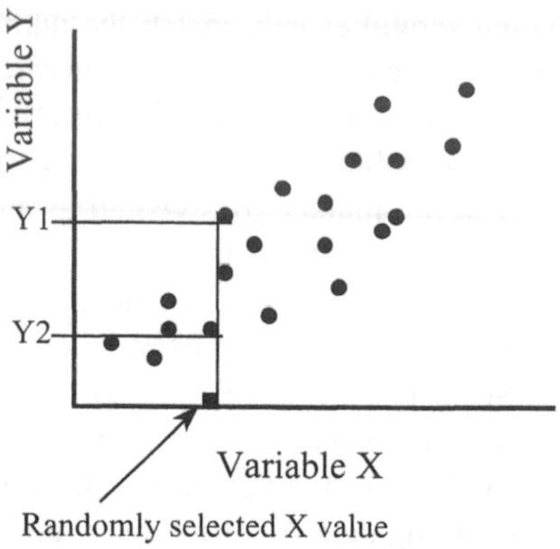

FIGURE 17.8 Cross plot of 2 variables with less-than-perfect correlation coefficient. Range Y1 to Y2 represents range of reasonable Y values for the randomly selected X value

sampled, and the value for the dependent variable will be randomly selected from a restricted range (such as that of Y1 and Y2 in Figure 17.8) within its distribution.

When historical data are not available and when you know that two variables are correlated, an educated guess will have to be taken as to the magnitude of the correlation coefficient. This is not as bad as it may sound. First, it is most important that two correlated variables are made to track one another. Often, the exact (if there is such a thing) calculation of a correlation coefficient is less important. In my classes I generally have students first assign a strong, moderate, or weak label to each variable pair that exhibits some correlation. A weak relationship can be represented by a correlation coefficient between something greater than 0 and ±0.55, a moderate relationship by a coefficient between ±0.55 and ±0.75, and a strong relationship by a coefficient of ±0.75 or greater. This certainly is not an exact science, but it generally serves the purpose.

MULTIPLE DEPENDENCIES AND CIRCULAR DEPENDENCE

In any given risk assessment there can be many pairs of variables that exhibit dependence. For example, the number of employees might correlate with the number of sick days taken (positive relationship — as one goes up, the other goes up) and the net income for our business unit may correlate with our costs/expenses (negative relationship — as costs go up, income goes down). It may also be true that one independent variable can have several associated dependent variables. Returning to our Stage of technology example, it may be that as an independent variable, Stage of technology may control several dependent variables. We already discussed the fact that at our corporation, as Stage of technology increases, our Customer commitment increases. This, then, is a positive correlation and would be indicated by a correlation coefficient between 0 and 1. However, it may also be true that our Stage of technology affects our corporation's cash exposure. That is, as we are increasingly viewed on Wall Street as the leader in this technology, it is

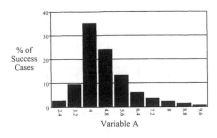 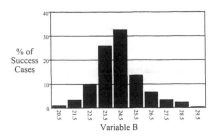

FIGURE 17.9 Frequency plot of independent variables A (left) and B (right).

increasingly easy for us to attract investors, thus lowering our corporation's cash investment. So, as Stage of technology increases, corporate cash exposure (investment) decreases. This is a negative relationship that would be represented by a correlation coefficient between 0 and −1. We can set up as many variables to be dependent on a single variable as we like. In this scenario, though, one must be careful not to set up circular dependence.

Any good software package should automatically guard against circular dependence. For example, for three variables, A, B, and C, circular dependence is indicated by: A is dependent on B, and B is dependent on C, and C is dependent on A. This circular logic should be rejected by the system. In this simple example, the circular reasoning is obvious. However, in complex technical and economic risk assessment models, it can be easy to inadvertently attempt to invoke such relationships.

EFFECT OF DEPENDENCE ON MONTE CARLO OUTPUT

Consider two independent (correlation of 0) distributions A and B from which we will calculate C by the formula C = A + B. The distribution for variable A ranges from 2 to 10 with a most likely value of 4. Distribution B ranges from 20 to 30 with a most likely value of 24. Both distributions are shown in Figure 17.9.

On a given Monte Carlo iteration, if an extreme value of A is selected (say, a value of 9), it most likely will be combined with a near-most likely value from B. The resulting cumulative frequency plot for C would look like that in Figure 17.10 and might range from a minimum C value of 23.7 to a maximum C value of 36.4.

If, however, a strong dependence of, say, 0.9 is established between A and B, then on any given Monte Carlo iteration when an extreme value of A is selected, dependence will force an extreme value of B to be combined with that of A. This will result in the plot of C shown in Figure 17.11 in which C now ranges from 22.4 to 40.0. A major effect, therefore, of implementing dependence is to change (generally expand) the range of the "answer" variable.

DEPENDENCE — IT'S UBIQUITOUS

Whenever I offer a risk class for managers, the class is given without computers. In a computerless environment, exercises must be designed to bring home a point without venturing too far into the erudite technical haze. One of my favorite com-

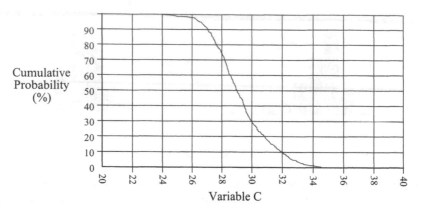

FIGURE 17.10 Cumulative frequency plot resulting from using Monte Carlo analysis to combine two independent variables A and B.

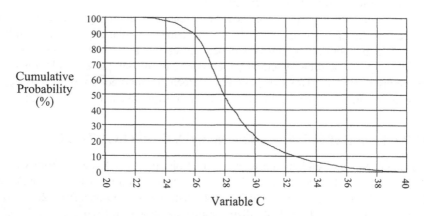

FIGURE 17.11 Cumulative frequency plot resulting from using Monte Carlo analysis to combine two highly correlated (0.9) variables A and B.

puter-free exercises is one I use to illustrate the ubiquitous nature of dependence and the need to consider it. Figure 17.12 is a list of variables from an actual plant-construction model. The original list was much longer. The list presented here has been whittled down for simplicity.

I typically challenge the class to peruse the list and then discuss which pairs of items on the list have real-world relationships (i.e., dependencies). I always get them kick-started by offering an example. The suggestion is put forth that Public perception and Litigation costs are linked. The reasoning is that if you are perceived as an "evil empire" by the public, you will spend more time in court. Next, I draw a line connecting these two items indicating that they have a relationship. As Public perception of your company worsens, Litigation costs go up. I then challenge the group to come up with some connections of their own. Typically, some lively discussion ensues.

Dependence

Public perceptions
Market research
Cost analysis
Production capacity
Demand
Profit margin per unit
Conversion costs (depreciated)
Conversion costs (actual)
Advertising costs
Litigation costs
Operating revenue
Tax rate
Marketing synergies
Taxable income

FIGURE 17.12 List of variables in dependence exercise.

The number of lines connecting variables in Figure 17.13 is about half of what can be drawn (I excluded the rest because it becomes too messy). As can be seen, there are many relationships, that is, dependent-variable pairs. In class, I project the original list on a white board or flipchart on which I draw the connecting lines. When the projector is turned off, you are left with just the tangle of lines shown in Figure 17.14 (remember, only about half the lines are shown). Class members generally are impressed concerning the number of relationships and by the fact that there is not one variable on the list that can be independently sampled — that is, sampled without consideration of at least one other variable.

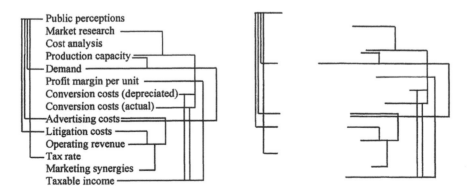

FIGURE 17.13 List of variables with dependencies shown as connecting lines.

FIGURE 17.14 Lines indicating dependencies.

I make the point that if you wish your risk model to emulate the real world, then the relationships shown must be honored. Even using such a simple list of variables, the relationships can be many and complex. If not considered, your risk model will be emulating something, but probably not what you had intended.

18 Risk-Weighted Values and Sensitivity Analysis

CONTENTS

INTRODUCTION TO RISK-WEIGHTED VALUES AND SENSITIVITY ANALYSIS

When a risk assessment is completed, interpretation of the results is the next step. Making this task as effortless as possible can be critical to the success of a risk model. If output from a risk assessment cannot be understood by those to whom it is presented, it likely was a wasted effort to have pursued the risk analysis in the first place. Risk-weighted values facilitate the integration of several risk-model-output components. The various types of risk-weighted values make simple the interpretation and comparison of output from multiple assessments.

Likewise, sensitivity analysis aids in identifying the elements of a risk model that were most and least important to the calculation of the answer. Following a risk-model run, examination of a sensitivity plot can be a fundamental step in deciding what action to take as a result of the analysis.

RISK-WEIGHTED VALUES — WHY?

The reason we need risk-weighted values always reminds me of the story of the dog chasing a car. The dog races down the street, expending copious quantities of energy, and finally catches the car. The question now becomes: "OK, buddy, you caught it — now what are you going to do with it?" So it is with risk assessments and their output. We struggle arduously to generate a detailed, comprehensive, and practical risk model and its resulting plots and erudite statistics. Well, OK, buddy, you got the results, now how are you going to interpret and compare output from multiple risk model runs?

299

It is useful to assess the magnitude of benefit or cost and the related probabilities associated with a single and isolated event. Relatively large returns, however, are realized only when we undertake the comparison of multiple scenarios or events. Without employing risk-weighted values, the process of comparison of risk-model outputs can be difficult and at times practically impossible.

Consider two investment opportunities. Opportunity A has a 20% chance of falling through. That is, if we invest in opportunity A, we have a 20% chance of losing our investment and realizing no return. If opportunity A does not fail, it will return no less than $3 million, no more than $12 million, and will most likely return about $5.3 million.

Like investment A, investment B has a chance of failure. There is a 35% chance that opportunity B will fail and our investment lost. If investment B does not fail, it will return a minimum of $1 million, a most likely of about $6.8 million, and a maximum of $10 million. Cumulative frequency plots for opportunities A and B appear in Figures 18.1 and 18.2, respectively.

Inspection of the plots in Figures 18.1 and 18.2 reveals that investment A has a 15% greater chance of succeeding than does investment B (20% vs. 35% chance of failure). If it does not fail, investment A has a higher minimum yield than does B. Likewise, A has a higher maximum yield. On the face of it, then, it might seem that A is the better opportunity; but there are other considerations.

From the cumulative frequency plot of investment B we can see that we have about a 45% chance of getting a return of around $6 million. The probability of realizing a $6 million return from A is less than half that of investment B. Perhaps we should take a look at this in a different space.

Figure 18.3 shows the frequency plots for the two opportunities. Inspection of the plots indicates that the most likely return from investment A, if it does not fail, is about $5.25 million. The most likely return from investment B, if it does not fail, is about $6.8 million. Well, now which investment appears to be better?

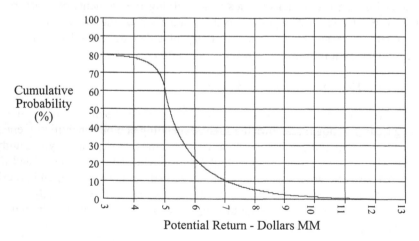

FIGURE 18.1 Cumulative frequency plot for Investment A.

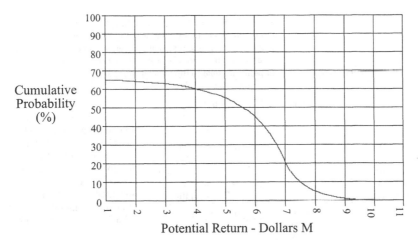

FIGURE 18.2   Cumulative frequency plot for Investment B.

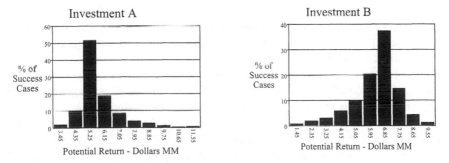

FIGURE 18.3   Frequency plots for two investment opportunities.

Risk-weighted values facilitate the comparison and ranking of probabilistically assessed opportunities. However, before we use risk-weighted values to select the best opportunity, let us take a look at just how risk-weighted values are calculated. The answer to this problem is given in a later section of this chapter entitled Risk-Weighted Values — The Answer.

RISK-WEIGHTED VALUES — HOW?

The fundamental premise behind the calculation of risk-weighted values is simple. In previous sections of this book we have elucidated on just how cumulative frequency plots are constructed (see Chapter 14). Figure 18.4 shows a set of values that make up a cumulative frequency curve. Figure 18.4 also shows an enlargement of the section of the cumulative frequency curve containing the two points whose coefficients are of greatest magnitude.

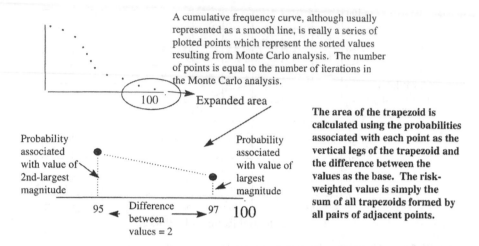

A cumulative frequency curve, although usually represented as a smooth line, is really a series of plotted points which represent the sorted values resulting from Monte Carlo analysis. The number of points is equal to the number of iterations in the Monte Carlo analysis.

100   Expanded area

The area of the trapezoid is calculated using the probabilities associated with each point as the vertical legs of the trapezoid and the difference between the values as the base. The risk-weighted value is simply the sum of all trapezoids formed by all pairs of adjacent points.

Probability associated with value of 2nd-largest magnitude

Probability associated with value of largest magnitude

95   Difference between values = 2   97   100

FIGURE 18.4 Plot showing how a risk-weighted value is calculated from a cumulative frequency curve.

The X-axis difference between the two points is 2 (99 – 97). The probability of random selection of the point whose coefficient is of greatest magnitude is one chance out of however many discrete X-axis values there are to choose from (the number of X-axis values is equal to the number of Monte Carlo iterations). Therefore, in the construction of the cumulative frequency plot, the "last" point is elevated one probability distance above the X axis. The Y axis is subdivided into equal-sized pieces. The number of pieces equals the number of Monte Carlo iterations. One piece is equal to one probability distance along the Y axis. The data value whose coefficient is second in magnitude is elevated two probability distances above the X axis (in a cumulative frequency plot, there are two chances out of the total number of discrete values of selecting the point of second-greatest magnitude *or* the point of greatest magnitude).

So, we know that the vertical legs of our trapezoid are probabilities and that the base of the trapezoid is the X-axis difference between the two points. We can now calculate the area of this trapezoid and all other trapezoids to the left of this one. The sum of all the individual trapezoidal areas is the risk-weighted value (we will see later that if the cumulative frequency plot reaches 100 on the Y axis, then the risk-weighted value calculated will equal the mean of the X-axis values).

Calculation of this area under the curve, however, is generally unsatisfactory for many reasons, not the least of which is the fact that the area remains constant regardless of where the curve is situated on the X axis. That is, the risk-weighted value for a curve with a minimum of, say, 10 and a maximum of 100 is exactly the same as if the same curve were "slid down" the X axis so that its minimum is 110 and its maximum is 200. The magnitude of the risk-weighted value does not reflect the position of the curve on the X axis. Therefore, some customized risk-weighted values generally are calculated. We will consider two of these.

THE NET RISK-WEIGHTED VALUE

No matter whether a risk assessment begins as a technical, political, environmental, or other type of evaluation, typically the result of such an assessment is a monetary or financial calculation. A most useful parameter for comparing and ranking financial risk model output is the net risk-weighted value.

In the previous section of this chapter it was explained how, in principle, a risk-weighted value is calculated and that calculation of only the area under the curve is not an intrinsically utilitarian measure. If risk-weighted values are to be used to compare and rank multiple risk assessed opportunities, then a mechanism must be employed whereby the changing position of the cumulative frequency curve, relative to the X axis, invokes a change in the risk-weighted value.

For the net risk-weighted value and the economic-risk-weighted-resource value (to be discussed in the next section of this chapter) the concept of the "limit line" is introduced. In Figure 18.5 a limit line has been selected that is coincident with the Y axis and the X-axis value of 0. The cumulative frequency curve in this figure is typical of many such curves which result from financial risk assessments in that the curve "straddles" 0; that is, the curve represents both positive (making money) and negative (losing money) values. X-axis values that plot far from 0 in either the positive or negative direction are of relatively great magnitude. The probabilities associated with these values, however, are relatively small. Therefore, on the negative side of the cumulative frequency curve, as with their positive counterparts, data points that plot increasingly far from 0 represent increasingly "large" negative values which also are increasingly less likely to happen in the real world (increasingly small associated probabilities).

As depicted in Figure 18.5, the positive area under the curve is calculated. This is done in the manner previously described. The negative area is calculated in exactly the same way (on the negative side, our trapezoids would be "upside down" with respect to their positive-curve counterparts). The negative area is then subtracted

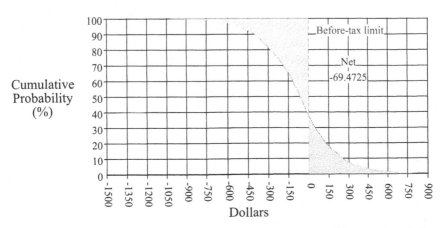

FIGURE 18.5 Cumulative frequency plot for 1999 before-tax annual increase in cash flow. Shaded areas are used to calculate net risk-weighted value.

from the positive area to yield a net risk-weighted area. This risk-weighted measure is a favorite tool of economists and others who must evaluate, compare, and rank financial opportunities. An example of its use will be given later in this chapter.

A net risk-weighted value, like any other risk-weighted value, is not an indication of how much an opportunity is expected to produce. That is, the risk-weighted value is not (necessarily) equal to the modal or most likely value on the answer-variable plot. Rather, a stand-alone, risk-weighted value indicates the dollar amount (or whatever the X-axis represents) that the risked opportunity would contribute to a portfolio of opportunities on a fully risk-weighted basis. The best use of a risk-weighted value, however, is as a means of comparison of one opportunity to the next. They are most commonly used as a relative measure. See the section Risk-Weighted Values — The Answer later in this chapter for an example.

THE ECONOMIC RISK-WEIGHTED RESOURCE VALUE

When the answer variable is expressed in other than monetary units such as gallons or barrels of liquid, tons, or other units that generally do not "go negative," the economic risk-weighted resource (ERWR) value may be employed. Like the net risk-weighted value, calculation of the ERWR utilizes a limit value — most commonly an economic limit.

Consider that we are in the midst of a labor strike. We have only enough non-union workers to operate one of our four factories. Figure 18.6 depicts a cumulative frequency curve which represents next month's projected production, in tons, from one of the factories. Values on the X axis indicate that next month's total production from the factory being considered for operation could range from 0 tons (if we, in fact, cannot make it work without union labor help) to 1000 tons. Our economists indicate that if total production is not at least 450 tons next month, then we will lose money for the month.

In response to our economist's calculated lower acceptable limit of production, we have drawn a limit line on the plot at 450 tons. The question we would like answered is: "What is the probability of achieving our lower limit of production, and what is the ERWR value (to be used in comparing one factory's projected output to that of the three others so we can decide which factory to operate)?"

An ERWR value is determined by calculating the area under the cumulative frequency curve "up to" the limit line and adding to this area the area of the "box" to the left of the limit line. The box extends all the way to the X-axis value of 0 whether or not it is shown on the plot. The shaded area in Figure 18.7 indicates the ERWR area. This, at first glance (or even at second glance), may seem to be an odd area to serve as the basis for a risk-weighted value.

Figure 18.8 is the frequency plot for the same data. The area represented by the bars to the right of the limit line in Figure 18.8 is the same area as that shaded in Figure 18.7. The area to the left of the limit line in Figure 18.8 is equivalent to the unshaded area in Figure 18.7 under the cumulative frequency curve. As with the net risk-weighted value, the ERWR is not the most likely amount the factory will produce next month (which is somewhere in the vicinity of the tallest bar in Figure 18.8). Rather, it is the amount of production that this factory would contribute to a portfolio

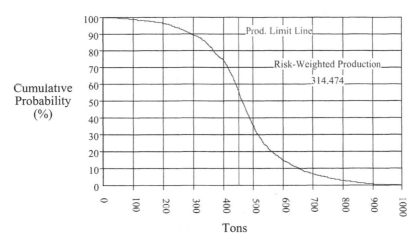

FIGURE 18.6 Cumulative frequency plot of projected production with limit line drawn at 450.

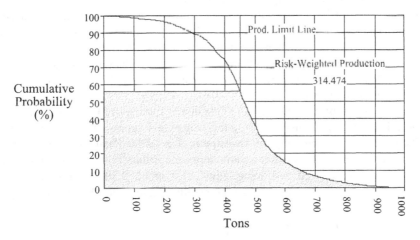

FIGURE 18.7 Cumulative frequency plot of projected production. Shaded area is used to calculate ERWR (risk-weighted production) value.

of outputs of factories on a fully risk-weighted basis. In the following section, an example is given which makes clear the use of risk-weighted values as ranking tools.

RISK-WEIGHTED VALUES — THE ANSWER

As previously indicated, although individual risk-weighted values do have meaning in the context of portfolio analysis, the real power of risk-weighted values is mainly realized when they are used to compare opportunities.

Let us return to the original problem presented in the Risk-Weighted Values — Why? section of this chapter. This problem presents us with the task of choosing

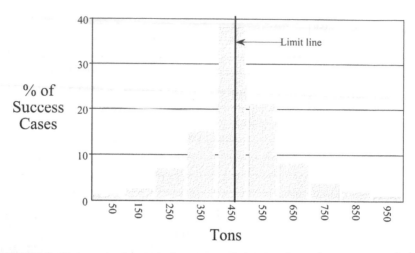

FIGURE 18.8   Frequency plot equivalent to cumulative frequency plot in Figure 18.7.

between two investment opportunities. Probabilistic analysis of the two opportunities yields the plots shown in Figures 18.1 and 18.2. The X axes of the cumulative frequency plots for both opportunities cover different ranges. The curve shapes are different. The Y-axis intercepts (chance of failure) of the two investments are not equal. Other parameters such as economic limits could be different (but are not in this example). How does a person mentally integrate changing ranges, changing curves, changing intercepts, etc.? Such integration is exactly the service provided by risk-weighted values.

Because investments generally are considered relative to 0, the limit lines used in the calculation of the net risk-weighted values in Figures 18.9 and 18.10 are at 0. The net risk-weighted value for investment A is 4.598. That for investment B is 4.125. Because the X axes of these plots represent dollars to be gained, investment A is the better choice (all else being equal, of course). If the X axis represented dollars to be paid, such as costs, then the smaller value would be the best choice.

SENSITIVITY ANALYSIS — WHY?

The purpose of most risk analyses is to ascertain the validity and quality of one or more opportunities. Most comprehensive risk studies are composed of many, some-times dozens, of input and output variables. These variables are linked by a system of equations that typically integrate technical, political, financial/commercial, and other parameters.

Each opportunity is assessed by the risk model with the expectation that output from the model will reveal the opportunity to be of good or poor quality — that is, an opportunity you might pursue or one you should forgo. If risk-model results indicate that a particular opportunity is of high quality, you should endeavor to determine which risk-model input parameters contribute most to its relative stellar standing. For example, it may turn out that an investment opportunity ranks high primarily because of the quality of the labor force (as opposed to favorable interest

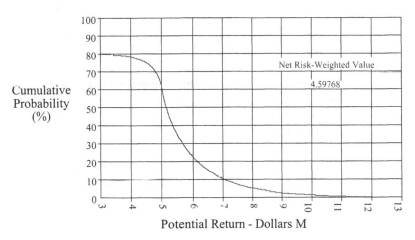

FIGURE 18.9 Cumulative frequency plot for Investment A and net risk-weighted value.

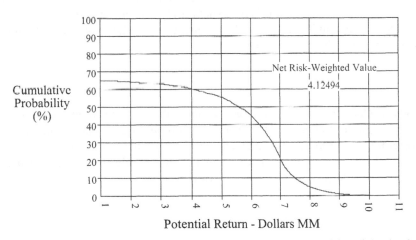

FIGURE 18.10 Cumulative frequency plot for Investment B and net risk-weighted value.

rates, quality raw materials, or other risk-model input parameters). Steps should then be taken to preserve the quality of the labor force so that the quality of the opportunity can be maintained. This is not to say that other parameters are not also important; however, generally only so much money, time, and effort can be directed toward any given opportunity. Therefore, it is essential to discover and nurture the most important contributing variables.

Conversely, the assessment of an opportunity may reveal it to be of poor quality. However, for political or other reasons we may wish to pursue the opportunity if its quality can be improved. A given opportunity may contain political, financial, environmental, commercial, technical, and other input variables. These parameters typically serve as the terms in complex equations that integrate them. If the quality of an opportunity is to be enhanced, it is essential to discover which parameters, if improved, would result in the greatest increase in quality of the opportunity. With limited resources it is beyond our ability to "fix" everything, so it is essential that we attempt to "fix" the parameters that will most improve the overall quality of the opportunity.

Inspection of the system of equations that comprise a risk model usually is of little or no help when attempting to identify critical variables. Unless the equations are simple sums or other equally elementary notation, perusal of the lines of computer code ordinarily will not reveal the magnitude of influence associated with individual variables. Neither can such influence, again with the exception of the most simple equations, be deduced from the magnitude of the coefficients that represent the variables. For example, if in a Monte Carlo–based risk model we consider the simple equation

$$C = A + B \tag{18.1}$$

in which A is a distribution with a minimum of 10,000,000 and a maximum of 100,000,000 while B is represented by coefficients that range from 2 to 6, it should be obvious that the value of C is overwhelmingly influenced by variable A. Variable B exhibits little influence. Variable A, therefore, is the "controlling" parameter. However, if we change the equation to:

$$C = A^B \tag{18.2}$$

now it is clear that the influence of B on the value of C has dramatically increased even though the magnitude of the coefficients representing A and B are unchanged. Therefore, except in the most mundane of circumstances, it is folly to suppose that the variables whose coefficients are of greatest magnitude also are those which exert the greatest influence on the "answer" variables.

SENSITIVITY ANALYSIS — HOW?

Through the years, many attempts have been made to develop algorithms which will reveal, for a complex set of equations, which of many input variables have greatest influence on the output variables. One of the more popular but significantly flawed methodologies is to hold all variables, except one, at a constant value in order to determine what influence the changeable variable has on the coefficients of the output variables. It should not take a rocket scientist, however, to realize that the true influence on an output variable depends on the interaction of all coefficient ranges for all variables and that the "hold-all-but-one-constant" (HABOC) method might not, in most cases, reveal the true influence exerted by any variable. Given the popularity of this approach, it is worth a few paragraphs here to describe some of the drawbacks to the HABOC method.

First, I would like to state that there is nothing intrinsically amiss with the form of a conventional tornado diagram (not shown here). If, for example, we have measured or calculated the ranges of net incomes of various businesses, there is not a thing wrong with displaying those net-income ranges as a tornado plot. The range in net income for each business can be represented by a bar of appropriate width and placed on the diagram accordingly. Such tornado plots are informative and ergonomically satisfying. So, the question concerning tornado diagrams is not whether they are truthful or useful, but rather how the information displayed was calculated.

In a risk model, the coefficients for all variables change, at least, with each iteration. It is difficult, by inspection, to ascertain the contribution of a single input variable to the change (relative to the previous iteration) in the value of an output variable. To remedy this, the HABOC methodology was born.

Using this approach, in order to determine the influence of a single input variable on an output variable, all variables, except one, are held at a constant value. The value used as the constant typically is a value determined to be "representative" of the distribution. Mean values, median values, and those termed *deterministic* commonly are employed. The Monte Carlo process is enjoined, holding all variables constant except one that is allowed to randomly vary throughout its range. Each variable has its turn to be allowed to vary while all others are held constant. For each variable allowed to change, the range of the output variable is noted, and this becomes the "width" of the bar for that input variable in the tornado plot. For speed (or for lack of sophistication), some approaches use only, for example, the 10th, 50th, and 90th percentile values for the variable that is allowed to change value. Discretization only worsens the problem.

The first thing that should strike you as suspicious concerning this methodology is that it violates dependence (see Chapter 17) and creates illegal combinations. For example, consider the two variables, Depth and Temperature, in a risk model. In the real geological world, these two variables are related. As depth (in the earth) increases, so does temperature (the deeper you go, the hotter it gets). This relationship is termed the *geothermal gradient* and varies with geographic location.

The Depth distribution might contain values ranging from 0 ft to 20,000 ft. Concomitant temperatures might range from 50°F to 500°F. Using the HABOC method, if you are attempting to determine the influence of Depth on an answer variable, you would allow Depth in the Monte Carlo process to vary throughout its 0 to 20,000 range while holding Temperature constant at its deterministic value. Because I have not here displayed either the Depth or the Temperature distributions, allow me to suggest that the deterministic value for Temperature is 220°F (it matters not what number you choose as the constant Temperature value — the result is the same).

Using the HABOC technique, Depth is allowed to vary while Temperature is held constant. So, Depth values of 100 ft are combined in the equations with the constant 220°F Temperature as are Depth values of 19,000 ft. These and countless other Depth/Temperature pairs are, as far as nature is concerned, illegal combinations. Yet, the variation of the answer variable is still, under this scheme, considered to reflect the influence of Depth on that variable. But this is not so in the real world.

Still another and more insidious drawback to the HABOC method is linked to branching in the risk-model logic. Consider the situation in which we have two variables A and B that are combined to create an output variable C. Variable A ranges, say, from 0 to 100. Consider also that our equations for calculating C contain a probabilistic branch thus:

$$\text{If } A < 50 \text{ then}$$
$$C = B + 1$$
$$\text{Else}$$
$$C = B^{100} \tag{18.3}$$

When attempting to determine the influence of B on the answer, the HABOC logic would have us hold A constant at some "deterministic" value. Again, it is irrelevant what value we select as the constant for variable A — let's pick 55. So, on each iteration the model will execute the If/Else logic and always calculate C to be equal to B^{100}. The C value of B + 1 will never be executed. The bar width indicating the influence of B on C will be drastically exaggerated because if A were allowed to vary, the equation B + 1 would be executed a statistically significant number of times. Therefore, the bar representing the influence of B on C is grossly in error. There is a better way.

To date, the most satisfying methodology developed for the purpose of discovering the relative influence of input variables upon the magnitude of coefficients for output variables is the Spearman Rank Correlation (SRC) method. In keeping with the practical bent of this book, a detailed description of the SRC equations was deemed unnecessary dunnage. Complete disclosure of the SRC methodology can be found in nearly any textbook on basic statistics (see Selected Readings for this chapter). However, it would be insoucient to refrain from putting forth at least a cursory description of Spearman's correlation method.

SRC is a nonparametric (distribution free) technique that facilitates the calculation of correlation while allowing all parameters in a multivariate analysis to vary simultaneously. Correspondence between individual input variables and a calculated output variable is arrived at by comparison of ranks. Ranks are employed in situations in which actual measurements are impractical or impossible to obtain. A mundane example is an ice-skating contest in which judges assign subjective scores. Such relative ranking methods are also used in many other activities in which absolute measurements are either impractical or impossible.

If measurements rather than ranks make up the original data set, then the measurements must be converted to ranks before SRC can be applied to the data. Conversion from measurements to ranks likely will result in the loss of some information. Therefore, it is unrealistic to expect that the product-moment correlation coefficient and that resulting from SRC will be equal when both methods are applied to a common set of data.

Loss of information notwithstanding, SRC may, relative to product-moment correlation, be a superior measure of correspondence. SRC's prelate position is derived from the fact that it is not predicated on the usually unwarranted presumption that the coefficients of the input variables are normally distributed. Correlation between variables representing coefficient distributions of nearly any type can receive valid treatment by the SRC technique.

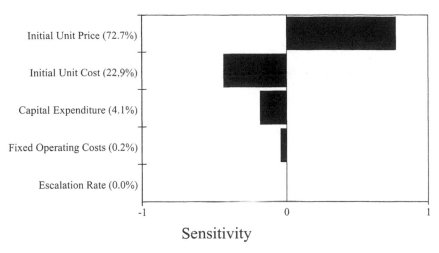

Sensitivity

FIGURE 18.11   Net income sensitivity plot.

SENSITIVITY ANALYSIS — RESULTS

Output from SRC analysis is commonly displayed as a sensitivity plot such as that shown in Figure 18.11. On a scale of −1 to +1, the length of the bars indicate the degree of correspondence between a given input variable and the calculated output variable. A separate sensitivity plot must be produced for each output parameter.

In the plot shown in Figure 18.11, the numbers in parentheses adjacent to the variable names indicate the percent of variance of the output variable that is explained or accounted for by that input variable. Bars extending to the right of the 0 center line are variables that are positively correlated with the output variable (as coefficients for the input variable increase in magnitude, so do the coefficients of the output variable). Bars extending to the left of the center line indicate negative correlations.

Sensitivity plots such as the one shown in Figure 18.11 not only are fundamental to determining which are the prominent input variables, but can also be invaluable indicators of whether or not a particular project should be pursued. For example, the results from a risk model may indicate that we likely will lose money if we pursue the risk assessed opportunity. For political reasons, however, we would like to go ahead with the project if the losses are not significant. We hope that our sensitivity analysis will reveal which of the aspects of the project we might seek to improve, thereby enhancing the palatability of the endeavor.

Our plot may indicate, however, that the two most significant variables, accounting for the majority of the variance of the output variable, are parameters over which we have little or no control. Parameters such as market price and taxes are good examples. As in this case, if parameters beyond our control are the significant factors in a discouraging risk assessment, then the sensitivity plot is not only an indicator of which variables are significant, but also is an indicator of whether it is logical and beneficial to further pursue an opportunity.

Selected Readings

INTRODUCTION THROUGH CHAPTER 11

Note: Because the introduction and first 11 chapters of this book relate to or are examples of risk/uncertainty analysis and, therefore, draw upon similar principles and precepts, the Introduction and Chapters 1 through 11 are here addressed as a single entity.

Goldberg, David E, *Genetic Algorithms in Search, Optimization and Machine Learning*, Addison-Wesley, Reading, MA, 1989.

Haupt, Randy L. and Haupt, Sue Ellen, *Practical Genetic Algorithms*, John Wiley & Sons, New York, 1998.

Koller, Glenn R., *Risk Assessment and Decision Making in Business and Industry: A Practical Guide*, CRC Press, Boca Raton, FL, 1999.

Malkiel, Burton, G., *A Random Walk Down Wall Street*, W. W. Norton, New York, 1990.

Markowitz, Harry M., *Portfolio Selection*, Blackwell, Cambridge, MA, 1991.

Rustem, Berc, *Algorithms for Nonlinear Programming and Multiple Objective Decisions (Wiley-Interscience Series in Systems and Optimization)*, John Wiley & Sons, New York, 1998.

Schrijver, Alexander, *Theory of Linear and Integer Programming (Wiley-Interscience Series in Discrete Mathematics and Optimization)*, John Wiley & Sons, New York, 1998.

Skinner, David C., *Introduction to Decision Analysis — Beginning Coursebook*. Probabilistic, 1996.

CHAPTER 12

Block, Peter, *The Empowered Manager: Positive Political Skills at Work*, Jossey-Bass Publishers, San Francisco, 1990.

Johnson, Barry, *Polarity Management — Identifying and Managing Unsolvable Problems*, HRD Press, Amherst, MA, 1992.

Koller, Glenn R., *Risk Assessment and Decision Making in Business and Industry: A Practical Guide*, CRC Press, Boca Raton, FL, 1999.

Maister, David H., *Managing the Professional Service Firm*, The Free Press, New York, 1993.

Morgan, Granger M. and Henrion, Max, *Uncertainty — A Guide to Dealing with Uncertainty in Quantitative Risk and Policy Analysis*, Cambridge University Press, Cambridge, MA, 1992.

Pascale, Richard Tanner, *Managing on the Edge — How the Smartest Companies Use Conflict to Stay Ahead*, Simon & Schuster, New York, 1990.

Skinner, David C., *Introduction to Decision Analysis — Beginning Coursebook*. Probabilistic, 1996.

CHAPTER 13

Skinner, David C., *Introduction to Decision Analysis — Beginning Coursebook*. Probabilistic, 1996.

CHAPTER 14

Bernstein, Peter L., *Against the Gods — The Remarkable Story of Risk*, John Wiley & Sons, New York, 1996.

Dore, A. G. and Sinding-Larsen, R., Editors, *Quantification and Prediction of Hydrocarbon Resources* from the Norwegian Petroleum Society (NPF), Special Publication No. 6, Elsevier, Amsterdam, 1996.

Meyer, Herbert A., Editor, *Symposium on Monte Carlo Methods*, John Wiley & Sons, New York, 1954.

CHAPTER 15

Davis, John C., *Statistics and Data Analysis in Geology, Vol. 2*, John Wiley & Sons, New York, 1986.

Ebdon, David, *Statistics in Geography — A Practical Approach*, Basil Blackwell, Oxford, U.K., 1981.

Harbaugh, John W., Davis, John C., and Wendebourg, Johannes, *Computing Risk for Oil Prospects: Principles and Programs*, Pergamonn, Tarrytown, NY, 1995.

Hayslett, H.R., *Statistics Made Simple*, Doubleday, New York, 1968.

Koller, Glenn R., *Risk Assessment and Decision Making in Business and Industry: A Practical Guide*, CRC Press, Boca Raton, FL, 1999.

Meyer, Mary A. and Booker, Jane M., *Eliciting and Analyzing Expert Judgement: A Practical Guide*, Academic Press, London, 1991.

Skinner, David C., *Introduction to Decision Analysis — Beginning Coursebook*. Probabilistic, 1996.

Stanley, L. T., *Practical Statistics for Petroleum Engineers*, Petroleum Publishing Company, Tulsa, OK, 1973.

CHAPTER 16

Nijhuis, H. J. and Baak, A. B., A Calibrated Prospect Appraisal System, *Proceedings of Indonesian Petroleum Association* (IPA 90-222), 19th Annual Convention, 69–83, October 1990.

CHAPTER 17

Campbell, John M. Jr., Campbell, John M. Sr., and Campbell, Robert A., *Analysis and Management of Petroleum Investments: Risk, Taxes, and Time*, CPS, Norman, OK, 1987.

Davis, John C., *Statistics and Data Analysis in Geology, Vol. 2*, John Wiley & Sons, New York, 1986.

Van Horne, James C., *Financial Management and Policy*, Prentice–Hall, Englewood Cliffs, NJ, 1974.

CHAPTER 18

Ebdon, David, *Statistics in Geography — A Practical Approach*, Basil Blackwell, Oxford, U.K., 1981.

Skinner, David C., *Introduction to Decision Analysis— Beginning Coursebook*. Probabilistic, 1996.

Index